Springer Theses

Recognizing Outstanding Ph.D. Research

For further volumes:
http://www.springer.com/series/8790

Aims and Scope

The series "Springer Theses" brings together a selection of the very best Ph.D. theses from around the world and across the physical sciences. Nominated and endorsed by two recognized specialists, each published volume has been selected for its scientific excellence and the high impact of its contents for the pertinent field of research. For greater accessibility to non-specialists, the published versions include an extended introduction, as well as a foreword by the student's supervisor explaining the special relevance of the work for the field. As a whole, the series will provide a valuable resource both for newcomers to the research fields described, and for other scientists seeking detailed background information on special questions. Finally, it provides an accredited documentation of the valuable contributions made by today's younger generation of scientists.

Theses are accepted into the series by invited nomination only and must fulfill all of the following criteria

- They must be written in good English.
- The topic should fall within the confines of Chemistry, Physics, Earth Sciences, Engineering and related interdisciplinary fields such as Materials, Nanoscience, Chemical Engineering, Complex Systems and Biophysics.
- The work reported in the thesis must represent a significant scientific advance.
- If the thesis includes previously published material, permission to reproduce this must be gained from the respective copyright holder.
- They must have been examined and passed during the 12 months prior to nomination.
- Each thesis should include a foreword by the supervisor outlining the significance of its content.
- The theses should have a clearly defined structure including an introduction accessible to scientists not expert in that particular field.

Peter L. Marek

Biomimetic Dye Aggregate Solar Cells

Doctoral Thesis accepted by
the Technical University of Darmstadt, Germany

Springer

Author
Dr. Peter L. Marek
Institute of Nanotechnology
Karlsruhe Institute of Technology
Karlsruhe
Germany

Supervisors
Prof. Dr. Teodor Silviu Balaban
Chirosciences Team
Aix-Marseille Université
Marseille
France

Prof. Dr.-Ing. Horst Hahn
Institute for Materials Science
Technical University Darmstadt
Darmstadt
Germany

ISSN 2190-5053 ISSN 2190-5061 (electronic)
ISBN 978-3-319-00635-2 ISBN 978-3-319-00636-9 (eBook)
DOI 10.1007/978-3-319-00636-9
Springer Cham Heidelberg New York Dordrecht London

Library of Congress Control Number: 2013938756

Printed on acid-free paper

Springer is part of Springer Science+Business Media (www.springer.com)

Parts of this thesis have been published in the following journal articles:

[1] Marek, P; Balaban, T.S., Biomimetische Solarzellen, *Nachr. Chem.*, **2006**, *54*, 1072-1077.

[2] Balaban, T.S.; Marek, P.; Hedderich, R., Solarzellen zur künstlichen Photosynthese nutzen Nanopartikel nach dem Vorbild grüner Schwefelbakterien, *Nanotechnik*, **2006**, *6*.

[3] Huijser, A.; Marek, P.L.; Savenije, T.J.; Siebbeles, L.D.A.; Scherer, T.; Hauschild, R.; Szmytkowski, J.; Kalt, H.; Hahn, H.; Balaban, T.S., Photosensitization of TiO_2 and SnO_2 by Artificial Self-Assembling Mimics of the Natural Chlorosomal Bacteriochlorophylls, *J. Phys. Chem. C*, **2007**, *111*, 11726-11733.

[4] Marek, P.L.; Sieger, H.; Scherer, T.; Hahn, H.; Balaban, T.S., Self-Assembled Chromophores Within Mesoporous Nanocrystalline TiO_2, Towards Biomimetic Solar Cells, *J. Nanosci. Nanotechnol.*, **2009**, *9*, 3708-3713.

[5] Marek, P.L.; Hahn, H.; Balaban, T.S., On the way to biomimetic dye aggregate solar cells, *Energy Environ. Sci.* **2011**, *4*, 2366-2378.

[6] Chappaz-Gillot, C.; Marek, P.L.; Blaive, B.J.; Canard, G.; Bürck, J.; Garab, G.; Hahn, H.; Jávorfi, T.; Kelemen, L.; Krupke, R.; Mössinger, D.; Ormos, P.; Reddy, C.M.; Roussel, C.; Steinbach, G.; Szabó, M.; Ulrich, A.S.; Vanthuyne, N.; Vijayaraghavan, A.; Zupcanova, A.; Balaban, T.S., Anisotropic Organization and Microscopic Manipulation of Self-Assembling Synthetic Porphyrin Microrods That Mimic Chlorosomes: Bacterial Light-Harvesting Systems, *J. Am. Chem. Soc.***2012**, *134*, 944-954.

Supervisor's Foreword

Peter L. Marek performed research at the Institute for Nanotechnology (INT, please see the list of Acronyms after the References) within the former Research Center (Forschungszentrum) Karlsruhe (FZK), now part of the Karlsruhe Institute of Technology (KIT) after unification with the University of Karlsruhe. He came to INT, first for a Diploma Thesis, which he submitted to the Technische Universität Darmstadt (TUD) and which was granted the highest mark. Encouraged by his mentor in Darmstadt, Prof. Dr.-Ing. Horst Hahn, the Managing Director of the INT, he joined the Balaban research group which was in the Department of Supramolecular Chemistry under the directorship of Prof. Dr., Dr. h. c. mult., Jean-Marie Lehn. Generous funding for assuring Peter's doctoral fellowship was provided by the FZK, while the Deutsche Forschungsgemeinschaft (DFG) provided extramural funding for equipment and consumables through the Center for Functional Nanostructures (CFN) at the University of Karlsruhe.

Our group at INT was performing mostly synthetic work on self-assembling chromophores which were the first fully synthetic molecules to mimic a very efficient natural light harvesting system of bacteriochlorophylls [1]. These are cousins of the very common plant chlorophylls, the most abundant pigments on earth. Peter, as trained material scientist at the TUD, embarked on an adventure which ended successfully with his Ph.D. defense, again judged with the highest mark, and with the present thesis, which you as reader are holding or reading on a screen, and which was under the joint supervision by Balaban and Hahn. Peter's research goal was to build solar cells which could function efficiently also under dim light, following the architectural principles of early photosynthetic bacteria. This biomimetic approach to solar cells was a novel endeavor specific to INT and FZK [2]. Peter brought to INT complementary knowledge and skills. While we provided the molecules through multi-step synthetic sequences, Peter studied their self-assembly by several different nanoscopic and spectroscopic techniques, providing unprecedented insight into their morphologies and their incorporation into solar cells. Peter is a team player with an extremely friendly personality and with whom unraveling unknown phenomena was exciting and stimulating. He has the ability to ask the right questions, which sometimes can be more important than searching for their answers.

The thesis is well written and, in my opinion, proves interdisciplinarity and a profound knowledge of solar cells with ample literature citations. The presentation is augmented by a very useful collection of procedures to build dye-sensitized solar cells either with liquid electrolytes or as all-solid state. Peter had the perseverance to surpass multiple obstacles, which in the end might affect the overall performance and long-time stability of solar cells. He is also an inventor and an admirer of bionics, which is a field imitating Nature from an engineer's perspective [3]. When something did not work, he looked for the reasons and tried to understand, improve, and surmount the difficulties. An example is the device shown in Fig. 6.4, which he designed and built in collaboration with specialists from the machine workshop of the FZK. This device allows electrical contacting for 16 solar cells assembled in parallel, on one and the same substrate, but which could have been differently treated or doped. The problem we initially encountered at the Linz Institute for Organic Solar Cells, was that upon contacting, in order to characterize solar cell performance, we short-circuited the sometimes brittle deposits [4]. Peter recognized this mechanical problem and brilliantly circumvented it. He was one of the most frequent users of the scanning electron microscopes at INT, as their log books show. The reader can convince herself/himself of the quality of Peter's images and his skill at finding the right information. Sifting through thousands of images, out of which only a selected few have been in the end incorporated into publications or this thesis, was a huge effort, requiring discipline, a rigorous archiving system, and expert computer technologies.

In collaboration with the group of Prof. Laurens Siebbeles and Dr. Tom Savenije at the Technical University in Delft, The Netherlands, we could measure by time-resolved microwave conductivity the exciton diffusion in wide band gap semiconductors such as nanocrystalline titanium dioxide [5]. A thorough understanding of the electron injection from the excited singlet state of chromophores, which behave collectively in such self-assemblies, into the conduction band of titania, was essential for optimizing cell dimensions and their nanomorphology. Furthermore, Peter could in collaboration with Dr. Ralph Krupke from INT perform an unprecedented pulsed dielectrophoretic deposition of self-assembled chromophores onto microstructured electrodes [6]. This work, while in the reviewing process, received a very favorable report stating that while for physicists it is quite common to have multiple authorship, chemists are often reluctant to publish work with more than 20 authors. Peter is the second author of this *J. Am. Chem. Soc.* paper, with 21 co-authors, but with equal contribution with the first author Cyril Chappaz-Gillot, who was a Ph.D. Student in Marseille, where Silviu Balaban now heads the *Chirosciences* Department of the Institute of Molecular Sciences of Marseille (iSm2). Cyril, a synthetic chemist educated at the "*Centrale Marseille*," earned the prize for the best fundamental science Ph.D. thesis of the *C'Nano* in the region Provence Alpes Côte d'Azur in France. It is only fitting that Peter, who had the hindsight to write his Ph.D. Thesis in English, earns the honor to publish a Springer Thesis. Mention should be made that Peter was awarded the prize for the best oral presentation "*On the Way towards Biomimetic Solar Cells*" *at the conference NanoSmat* in Porto (Portugal) in 2007.

Peter often made presentations for high-school students on solar cells, sometimes with practical demonstrations of sensitizing titania with wild berry fruit juices. His vivid talks always left a "begeistert" audience. If we do not start now to educate the next generation of potential researchers on how to solve the conversion of the rather dilute solar energy to concentrated electricity or fuels, by mid-century, our civilization will face a huge energy problem, especially if population-rich Asian or Latin American countries reach the standard lifestyle of Europe, North America or Australia and New Zealand. As written by Vicenzo Balzani, the creator of the Bologna School on Photochemistry at Giacomo Ciamician's University: "The efficient production of solar fuels would represent the most important breakthrough of modern science" [7].

One anecdotal recollection from Peter's early days at INT, was when one morning he very boldly put on a brilliant white new lab coat, upon my insisting that he perform a column chromatography on his 'workhorse compound **P2**,' as entitled in this thesis, which was formerly repeatedly synthesized by Anil Bhise, Dennis Mössinger, Chilla Malla Reddy, and more recently by Cyril. After explaining how chromatography works, showing Peter how the four rotary evaporators were functioning and where **P2** was migrating on his column, I left him alone for the rest of the day. In the evening, a radiant Peter, in a dirty lab coat, completely exhausted, showed me the powdered compound in a flask under argon. He acknowledged this had been physically, his most exhausting day at INT. The result of my brutal throwing a non-swimmer into the pool and waiting for him to discover how to swim, which in Peter's case translated into demanding from a non-trained chemist to prepare his object of study, was that Peter gained a deep respect for his beloved compound. It often happens that interdisciplinary work between chemists and physicists or materials scientists is hampered, less than by each community's jargon or usage of different units for energy than by the fact that physicists are using complicated instruments which need tuning, and several months of assembly until experiments can be meaningfully performed. Once such a sophisticated experiment is up-and-running, physicists are in need of samples. Biologists, Biochemists or Nature are favored sample sources, in addition to commercial suppliers. However, it is often the chemist, who has an exact knowledge of how to handle the compound he synthesized, sometimes also in month-long synthetic sequences. The chemist knows his solvents and how to ultrapurify these. He can avoid adventitious impurities, such as either the stabilizer of the commercial solvent, or the phthalate plasticizer from the plastic cap of his vial that can drastically quench photoexcited states. The perfect marriage is when a pure sample, in an ultra-pure medium, is put in a well-tuned instrument, allowing new knowledge to be gained, reproducibly. Peter learned this, probably the hard way, and just as in Tango he took two steps back, two steps in front, but having fun most of the time.

Another remark I would like to reiterate here from a previous Account [8] is that "Man-made devices are often better than natural ones for performance and robustness. Typical examples are wheels and cogwheels, for which there exists no natural counterpart, but when coupled to an engine allow much higher velocities than any natural joint designed for movement. However, for light-harvesting one still awaits solutions. Learning from Nature's architectural principles might help us to devise efficient devices with robust pigments allowing functioning at elevated temperatures (over 80 °C in the deserts), for long periods (desirable are over 20 years of serviceless functioning), or under low-light illumination conditions."

Peter's work, described in the following pages, is an effort in this direction. Hopefully, this work will stimulate others to join in what we think is a worthwhile and stringently needed endeavor.

From another paper co-authored with Peter and Horst I cite here: "Organic and hybrid solar cell research is still an emerging field and is in a development stage where organic light emitting diodes were a decade ago. The problems then encountered with the stability of displays have been solved in the course of the past few years so that flat displays based on organics are now commonplace. We consider that similar breakthroughs can be achieved for the solar cell field. Ideally, hybrid solar cells functioning with over 10 % power conversion efficiencies also under dim or diffuse illumination conditions should be attainable" [9].

Finally, citing from Giacomo Ciamician's (father of photochemistry and a bionic's pioneer) landmark *Science* paper in 1912: "On the arid lands there will spring up industrial colonies without smoke and without smokestacks; forests of glass tubes will extend over the plants and glass buildings will rise everywhere; inside of these will take place the photochemical process that hitherto have been the guarded secret of the plants, but that will have been mastered by human industry which will know how to make them bear even more abundant fruit than nature, for nature is not in a hurry and mankind is" [10].

Acknowledgment: Dr. Tobias Wassermann from Springer is thanked for his initiative regarding Peter's thesis, for his careful proofreading, and patience for the final manuscript.

Marseille, April 2013 Prof. Dr. Teodor Silviu Balaban

References

1. Balaban TS, Bhise AD, Fischer M, Linke-Schaetzel M, Roussel C, Vanthuyne N (2003) Controlling chirality and optical properties of artificial antenna systems with self-assembling porphyrins. Angew Chem Int Ed 42:2139–2144; 115:2189–2194 (Frontispiece illustration). doi:10.1002/anie.200250465
2. Bhise AD (2004) A biomimetic approach for synthesizing artificial light-harvesting systems using self-assembly, Wissenschaftliche Berichte, FZKA 7174, Forschungszentrum Karslruhe in der Helmholtz-Gemeinschaft, 2005, ISSN 0947-8620. Ph.D. Defense on December 14, 2004

3. Wikipedia definition of bionics (also known as biomimicry, biomimetics, bio-inspiration, biognosis, and close to bionical creativity engineering) is: "The application of biological methods and systems found in nature to the study and design of engineering systems and modern technology". Accessed 9 Nov 2012
4. Arici E, Sariciftci NS, Balaban TS (unpublished results)
5. Huijser A, Marek PL, Savenije TJ, Siebbeles LDA, Scherer T, Hauschild R, Szmytkowski J, Kalt H, Hahn H, Balaban TS (2007) Photosenzitization of TiO_2 and SnO_2 by artificial self-assembling mimics of the natural chlorosomal bacteriochlorophylls. J Phys Chem C 111:11726–11733. doi:10.1021/jp072927s
6. Chappaz-Gillot C, Marek PL, Blaive BJ, Canard G, Bürck J, Garab G, Hahn H, Jávorfi T, Kelemen L, Krupke R, Mössinger D, Ormos P, Malla Reddy C, Roussel C, Steinbach G, Szabó M, Ulrich AS, Vanthuyne N, Vijayaraghavan A, Zupcanova A, Balaban TS (2012) Anisotropic organization and microscopic manipulation of self-assembling synthetic porphyrin microrods that mimic chlorosomes: bacterial light-harvesting systems. J Am Chem Soc 134:944–954. doi: 10.1021/ja203838p
7. Armaroli N, Balzani V (2007) The future of energy supply: challenges and opportunities. Angew Chem Int Ed 46:52–66
8. Balaban TS (2005) Tailoring porphyrins and chlorins for self-assembly in biomimetic artificial antenna systems. Acc Chem Res 38:612–623. doi: 10.1021/ar040211z (Front cover illustration)
9. Marek PL, Hahn H and Balaban TS (2011) On the way to biomimetic dye aggregate solar cells. Energy Environ Sci 4:2366–2378; (Perspective Article). doi:10.1039/c1ee01053k
10. Ciamician G (1912) The photochemistry of the future. Science 36:385–394. Cited after an inspiring talk by Thomas Moore, Arizona State University, in Montreal at the 13th International Congress of Photosynthesis, 2004

Acronyms

CFN	Center for Functional Nanostructures, a DFG excellence cluster, formerly at the University of Karlsruhe (TH), since 2010 at the Karlsruhe Institute of Technology
DFG	Deutsche Forschungsgemeinschaft, for which the French equivalent is the ANR, Agence Nationale de la Recherche
FZK	Forschungszentrum Karlsruhe; formerly known as KFZ (Kernforschungszentrum, or Nuclear Research Center), now KIT, campus North
INT	Institut für Nanotechnologie, the first Nanotechnology Institute founded in 1998; The Balaban Laboratory started in a container building in August 1999
iSm2	Institut des Sciences Moléculaires de Marseille, the present research body harbouring the Balaban lab
KIT	Karlsruhe Institute of Technology, a unique research and teaching facility in the German science landscape comprising the Max-Planck Institutes, the Hermann von Helmholtz Association and the Fraunhofer Institutes together with the Universities. KIT is a merger of the FZK and the University of Karlsruhe (TH)
TUD	Technische Universität Darmstadt

Acknowledgments

I thank Silviu Balaban for giving me the opportunity to work on this highly interesting topic. I very appreciate all the support and time he spend for educating me in far more fields than chemistry and biology. He has been always a fair, friendly, broad-minded and competent supervisor with an open and human manner. Thank you also for the great support in writing publications and the nice time in your group.

I thank Horst Hahn for giving me the opportunity to work in the Nice Institute for Nanotechnology, for his good ideas and advices as well as for the careful corrections of my thesis and the financial support. I very appreciate his human manner and also the nice events and workshops he organized for us, like in La Clusaz.

I thank also Prof. Wolfram Jaegermann for raising my interest in the field of solar cells and giving me the opportunity to use his facilities, as well as for being the referee for my thesis. I thank also Eddie Feldmeier, Konrad Schwanitz and Thomas Mayer for their good advices and help to characterize my solar cells.

I thank Prof. Uli Lemmer and his group, in particular Alexander Colsmann and Michael Klein for their help in characterizing my solar cells and the experiments incorporating our dye P2-diacetyl into organic solar cells. Many thanks go also to Anatoliy Slobodskyy for the nice collaboration characterizing my DASCs and his CIGS solar cells and for the aluminum doped zinc oxide substrates for my tests.

I want to thank also Prof. Heinz Kalt and his group, in particular Robert Hauschild, Jędrzej Szmytkowski and Jonas Conradt for the good collaboration, the time-resolved fluorescence measurements and the zinc oxide nanorod substrates.

Tom Savenije and Annemarie Huijser I want to thank for the nice collaboration and for the professional time-resolved microwave conductivity measurements.

I also thank Clare Dyer-Smith for the interesting discussions and the transient absorption spectroscopy measurements. Dadhichi Paretkar and Oliver Perroud I want to thank for the nice time in Bangalore and the stuff they send me via email.

My thanks go also to Ralph Krupke and Aravind Vijayaraghavan for the good collaboration and the extensive dielectrophoretic deposition experiments and all the help with the SEM.

I want to thank also Christian Kübel for the professional TEM images, Wiebke Lohstroh for the TGA, Malla Reddy for the porphyrin synthesis and the help in the lab. Sebastien Vanneste I thank beside his frequent help in the chemistry lab especially for making this work so much enjoyable and also for the lessons in climbing. Carmen Balaban I thank for managing everything for us in the chemistry lab and the familiar atmosphere in the work group and all the help and good advices.

Thank you Babak Nasr, Raghav Wahi and Simon Bubel for your generous help and the nice time I could spend with you in the physics labs and in my free time.

Torsten Roßmann I thank for the nice bionics lessons and for giving me the opportunity to hold a lesson in the bionics "Ringvorlesung" about the results of my Diploma Thesis.

I thank the INT football team, in particular our great coach Matthias Hettler, not only for the nice weekly sessions on the field and in the Andreas Bräu.

My thanks go also to the rest of the INT for the kind atmosphere and to the others, who contributed more or less to the success of this work and who I forgot to mention in particular.

My special thanks go to Torsten Scherer, for being a great college and friend, helping me a lot to obtain professional results with many devices, especially SEM and EDX, and for making the work so joyful. Thank you for the professional FIB and TEM measurements and for merging our colleges together and making our institute friendlier, greener and funnier, for example by organizing the nanocinema events. Without our discussions during cycling every day or our adventures climbing the Etna or in the Alps or just relaxing with nice movies after long workdays, the present work would have become by far not this good.

Many thanks go also to Bert Kempf for his generous and professional help with scientific discussions and corrections of my English grammar. I'd like to thank also Gudrun and Elisabeth for their warm hospitality and the nice time to relax.

And of course I thank my family for giving me a good education and supporting me wherever possible.

I am very thankful for heaving had Sonja by my side almost all the time, motivating me to do this work, being so patient with me and giving me a beautiful time.

And last but not least I want to thank also Anja Maier for helping me to finish this work and to prepare myself for the defense and for being such a great woman.

Contents

Abbreviations

Physical Abbreviations	Description
absorbance	In this thesis: decadic absorbance (A_{10})
APCE	Absorbed photon to current (conversion) efficiency
a-Si	Amorphous silicon
BLH	Biomimetic light-harvesting
BChl *c*	Bacteriochlorophyll *c*
c-Si	Crystalline silicon
DASC	Dye aggregate solar cell
DSSC	Dye sensitized solar cell
DEP	Dielectrophoresis
E_F	Fermi energy
EQE	External quantum efficiency
FF	Fill factor
HOMO	Highest occupied molecular orbital
HTM	Hole transport material
IDE	Interdigitated electrode
IPCE	Incident photon to current (conversion) efficiency
IQE	Internal quantum efficiency
I_{MPP}	Current at the maximum power point
I_{SC}	Short circuit current
LUMO	Lowest unoccupied molecular orbital
LHS	Light-harvesting system
Mpp	Maximum power point
P	Power
PES	Photoelectron spectroscopy
R_S	Series resistance
R_{sh}	Shunt resistance
SEM	Scanning electron microscope
SPD	Spray pyrolysis deposition
ss-DSSC	Solid-state dye sensitized solar cell
TEM	Transmission electron microscope
UHV	Ultrahigh vacuum

V_{MPP}	Voltage at the maximum power point
V_{OC}	Open circuit voltage
XRD	X-ray powder diffraction

Materials and Compounds	**Description**
DCM	Dichloromethane (CH_2Cl_2)
diethyl ether	$C_4H_{10}O$
FTO	SnO_2:F, Fluorine-doped tin oxide
ITO	In_2O_3:Sn, Indium tin oxide, (tin-doped indium oxide)
N3	Ru(dcbpyH_2)$_2$(NCS)$_2$, dcbpyH_2 = 4,4′-dicarboxy-2,2′-bipyridine
N719	(Bu_4N)$_2$[Ru(dcbpyH)$_2$(NCS)$_2$], dcbpyH = 4,4′-dicarboxy-2,2′-bipyridine
n-heptane	*normal*-heptane (C_7H_9)
o-DCB	*otho*-dichlorobenzene ($C_6H_4Cl_2$)
P2	3-(1-hydroxyethyl)-13-acetyl-10,20-bis-(3,5-di-*t*-butylphenyl) zinc porphyrin
P2-diacetyl	3,13-acetyl-10,20-bis-(3,5-di-*t*-butylphenyl) zinc porphyrin
P3HT	Poly(3-hexylthiophene)
PBT	Thieno[3,4-*b*]thiophene-benzodithiophene
PCBM	[6,6]-phenyl-C_{61}-butyric acid methyl ester
PEDOT:PSS	Poly(3,4-ethylenedioxythiophene):poly (styrenesulfonate)
Sb dopant	N(PhBr)$_3$SbCl_6, (tris (4-bromophenyl)ammoniumyl hexachloroantimonate
tBP	4-tert-butylpyridine
spiro-MeOTAD	2,2′,7,7′-tetrakis-(*N*,*N*-di-*p*-methoxyphenyl-amine)9,9′-spirobifluorene
TFSI-Li	Li[CF_3SO_2]$_2$N, bis(trifluoromethane)sulfonimide lithium salt
THF	Tetrahydrofuran (C_4H_8O)
Ti-iProp.	Titanium isopropoxide ($C_{12}H_{28}O_4Ti$)
TiO_2	Titanium dioxide, titania

Chapter 1
Introduction

The solar cell architecture, developed in this work, takes advantage of a combination of hybrid solar cells and solid-state dye sensitized solar cells (ss-DSSCs), also called solid-state dye solar cells (SDSCs). Hybrid solar cells are based on a junction between an inorganic and an organic semiconductor, whereas the latter one is typically supposed to absorb the light, due to its higher absorption coefficient. This organic semiconductor has the additional duty to transport the light excitation energy in form of an electron hole pair (exciton) to the inorganic semiconductor, inject there an excited electron and transport the hole to the opposite side of the cell. Therefore, it ideally should have a high absorption coefficient and a high mobility and lifetime for the excitons and holes. As such organic all-rounders are rare and have not reached high efficiencies, yet, the ss-DSSC pursues another strategy: the separation of the functions onto different materials. Whereas the inorganic semiconductor keeps its function for transporting the electrons, the dye in DSSCs has no ability to transport excitons and holes. This is taken over by an additional organic material, the so-called: hole transport material (HTM). However, this is the other extreme, because the dye has no ability to transport excitons or charges. Consequently it needs a large area which controls its assembly to a monolayer by chemisorption in order to allow the contact of each dye molecule from two sides by different materials. This area is typically provided by a mesoporous TiO_2 layer, which is covered by the dye monolayer and infiltrated by the HTM. Any aggregate formation of these dyes needs to be prevented and ideally also the random contacts between the dye molecules within the monolayer, because this leads to a fast quenching of the excitons. For an efficient exciton transport the dyes need to be orderly assembled, which can be achieved by a sophisticated functionalization for self-assembly. How to achieve this could be learned from natural early life forms, like bacteria, which made much more use of self-assembly compared to higher developed organisms like plants, which mostly employ proteins to assemble functional structures. Luckily some of these archaic species, like the photosynthetic active green sulfur bacteria, sustained their position in inhospitable niche regions, like near hot volcanic springs or in regions with the lowest light intensities on earth. Their light-harvesting system, also called

P. L. Marek, *Biomimetic Dye Aggregate Solar Cells*,
Springer Theses, DOI: 10.1007/978-3-319-00636-9_1,

antenna system, demonstrates the potential of self-assembly to form highly ordered dye aggregates, which allow for an efficient exciton transport, and thus an efficient energy conversion in a smaller volume than needed for dye monolayers. Transferring this principle of self-assembly to artificial dyes might pave the way to improve conventional solar cells.

Light absorption is the first step of solar energy conversion. As the energy of solar radiation, i.e., the product of the number of photons and their energy, is maximal in the visible wavelengths range, all efficient solar cells absorb in this spectral region and thus are colored. Beside the intrinsically colored semiconductors, like silicon or the compound semiconductors for thin film solar cells, it is also possible to use optically transparent inorganic or organic semiconductors to construct solar cells by staining them with dyes. These so called dye sensitized solar cells (DSSCs) are typically based on the high bandgap semiconductor TiO_2, which is stained by a monolayer of an organic dye. The dye sensitizes the TiO_2 for visible light, whereas otherwise the TiO_2 would be only able to absorb and convert ultraviolet light to electricity. As organic dyes reach higher extinction coefficients as inorganic ones and are potentially cheap in manufacture, they are preferred candidates for low cost solar cells. Their current disadvantage of a reduced lifetime, reaching a couple of years, does not make them attractive for mass energy production, yet, but they have good chances to capture the market in niche applications like low cost solar cells for consumer electronics. There the lifetime of the solar cell is of less importance, because such electronics, like mobile phones or notebooks, are not intended for a many years long use, anyway. Additionally, such solar cells may also be replaced together with the battery, which is also accepted to have a life time limited to a couple of years. Furthermore the energy needed to manufacture solar cells based on organic dyes is much smaller than that needed for conventional inorganic solar cells. During the short lifetime of consumer electronics and in the weak light illumination conditions conventional solar cells might not be able to pay back the energy, which was used for their manufacturing. Thus, solar cells based on organic dyes are potentially better suited for consumer electronics and inorganic solar cells for roof-top applications.

Most conventional organic dyes can be applied to solar cells only in form of a monolayer, because their aggregates or multilayers do not reach a degree of ordering, necessary for an efficient energy transport over several dye molecules. To accommodate enough dye in form of a monolayer to absorb the full sunlight intensity, a large surface area is required. A high ratio of surface area to volume can be achieved with a very fine porous network like the mesoporous TiO_2 used in DSSCs. As liquid electrolytes easily penetrate all pores and form a beneficial barrier preventing recombination at the interface to the TiO_2, such DSSCs achieve the highest efficiencies of all solar cells with organic components. However, the liquid electrolyte has certain disadvantages, like complicating the cell assembly and decreasing its long-term stability, due to sealing and leakage problems. Additionally, these solar cells need to be rigid, what increases their weight. Hence, the research effort to find solid-state alternatives for this electrolyte steadily increased within the last decade. However, not many solid (HTMs) exist, which could infiltrate the small mesopores. Additionally, the recombination rate between the HTM and the TiO_2 is around hundred times higher than with the liquid electrolyte. It was concluded therefore, that a more coarse-porous TiO_2 structure would help to solve these problems, as it increases the pore size on one hand to facilitate the HTM infiltration and decreases the recombination affected surface/ interface area on the other hand. This would be only possible in combination with dye multilayers instead of monolayers to accommodate enough dye on a smaller surface area. However, dye aggregates are only able to transport their excitation energy efficiently in form of excitons, when they are highly regular arranged next to another. As only very few dyes have been developed, which are able to self-assemble to such structures, Silviu Balaban examined the possibility to adopt the natural self-assembly principle to artificial dyes [1].

In this thesis, a biomimetic self-assembling dye, based on a zinc porphyrin, was investigated toward its suitability and applicability for solar cells. Previous investigations on this porphyrin showed the possibility to control the aggregate morphology and size to a certain degree within porous substrates by a two step process [2, 3]. Thereby, the porphyrin was first infiltrated from a solution in tetrahydrofuran (THF) into a mesoporous TiO_2 layer containing additional macropores. The prevented self-assembly in this solvent allowed for an efficient infiltration, but led to the formation of amorphous aggregates. As this would not allow for efficient solar cells, a post-treatment in anhydrous *n*-heptane has been developed for a reorganization of the aggregates to the desired well-defined aggregates.

The aim of the present thesis was to better control the aggregate morphology and size in order to achieve a narrow distribution of these aggregate structures, and hence optoelectronic properties, and their micro- and macroscopic homogeneous implementation into solar cells. This thesis is therefore divided into two parts: the first one deals mainly with investigations on how the self-assembly could be controlled in order to achieve a homogeneous aggregate deposition with suitable optoelectronic properties. The second part is about the conception and construction of solar cell prototypes and their characterization.

1.1 Terms and Definitions

Aggregate, self-assembly, crystal. Aggregate is a generic term for solids regardless of their structure, which can range from a completely disordered (amorphous) one up to a highly regular ordered crystal. With the term self-assembly commonly a degree of spontaneous ordering is meant, which has a sophisticated growth mechanism or error correction mechanism during the crystal growth, which is often encountered in primitive natural organisms [4]. This may combine a fast growth rate with a low density of defects, due to an error correction mechanism, which effectively inhibits the further growth onto a misplaced unit. Self-assemblies combine more or less crystal properties, like a high degree of order, but typically with a less extended long-range order, with additional properties such as an increased mechanical flexibility or tolerance for changes of the environmental conditions during the growth process, like the temperature.

Chromophore, pigment, dye. A chromophore (Greek: color-bearer) is originally the part of a molecule responsible for the color, but this term is commonly also used in a broader sense for relatively small entire molecules, like chlorophylls. These molecules may also be called pigments, though this term originally meant insoluble aggregates or particles. Dyes in the original sense are soluble chromophores with a specific functionality mostly for chemisorption onto certain materials, but this term is also used in analogy to chromophores.

Exciton. An exciton is an electron-hole pair, which originates typically by the absorption of light in organic materials. Whereas in inorganic semiconductors the excited electron is weakly bound to its remaining hole and dissociates very fast into separated charges, the binding energy of excitons in organics is much higher, typically between 0.5 and 1eV [5]. Thus, excitons in organics normally dissociate only at suited interfaces to other materials, so-called hetero-junctions. Excitons can move between chromophores by the so-called Förster resonance energy transfer (FRET) or the Dexter energy transfer mechanism.

Mesopore. Mesopores are pores of an intermediate size between 2 and 50 nm. Smaller pores are called micropores and larger ones macropores [6]. This historic naming might be confusing as micropores might be associated with micrometer sized pores, what is not the case by this definition. The term nanopore is also used sometimes, instead of mesopore.

References

1. Balaban TS (2005) Acc Chem Res 38:612–623
2. Marek PL, Sieger H, Scherer T, Hahn H, Balaban TS (2009) J Nanosci Nanotechnol 9:3708–3713
3. Marek P (2005) Selbstanordnende Zn-Porphyrine in makroporösen TiO_2-Schichten für farbstoffsensibilisierte Solarzellen. Diploma thesis, Technical University of Darmstadt
4. Eigen M (1971) Naturwissenschaften 58:465–523
5. Brütting W (2005) Physics of organic semiconductors. Wiley-VCH, Weinheim
6. McNaught AD, Wilkinson A (1997) IUPAC. Compendium of Chemical Terminology, 2nd edn. (the "Gold Book"). Blackwell Scientific Publications, Oxford. XML on-line corrected version created by Nic M, Jirat J, Kosata B; updates compiled by Jenkins A: http://goldbook.iupac.org. doi:10.1351/goldbook, doi:10.1351/goldbook.M03853, ISBN: 0-9678550-9-8

Part I
Biomimetic Light-Harvesting

Chapter 2
Theory and Literature Survey: Biomimetic Light-Harvesting

2.1 Natural Light-Harvesting

The reaction centers, where charge separation for chemical energy storage occurs, are almost identical for all photosynthetic organisms. This fact hints at the same ancestor [1]. In contrast, the light-harvesting systems or antenna systems can have very different architectures from species to species as they are adapted to specific habitats and illumination conditions. The antenna systems of plants, for example, are complicated chromophore-protein complexes, whereas those of green sulfur bacteria consist of self-assembled chromophores. However, the basic principle is the same for all antenna systems: the exciton coupling. It ensures within precisely arranged chromophores an efficient exciton transport within the antenna system toward the reaction center. Therefore, only one reaction center is needed to convert the absorbed light energy of many chromophores, what results in a very economic light-harvesting strategy. Additionally, the wavelength range, which can be absorbed by single chromophores, increases due to the exciton coupling of the well arranged chromophores, what allows to achieve higher energy conversion efficiencies [2].

If the dye molecules are not highly regular oriented next to each other, the exciton transfer between them is very inefficient and they get quenched very fast. As artificial proteins, which maintain the arrangement of plant chlorophylls for the excitonic coupling, cannot be synthesized in a large scale, it is worth taking a look on nature to study how earlier lifeforms achieve similar orderly structures without the help of proteins. Green sulfur bacteria for instance, contain the self-assembling bacteriochlorophylls (BChls) *c*, *d*, or *e* (Fig. 2.4). They are supposed to self-assemble to the largest and most efficient light-harvesting antennae found in nature [3]. Each antenna system forms the center of an organelle, which is called chlorosome (Greek: green sac). It contains a so-called baseplate, through which the excitons are transferred to the reaction center. The chlorosomes reach dimensions of roughly: $150 \times 50 \times 20$ nm, which vary between the species as they are adapted to different illumination conditions.

P. L. Marek, *Biomimetic Dye Aggregate Solar Cells*,
Springer Theses, DOI: 10.1007/978-3-319-00636-9_2,

Green sulfur bacteria live in waters with a very low oxygen content and in the presence of hydrogen sulfide, which they use for photosynthetic cleavage instead of water. They are found in hot springs close to the water surface (*Chlorobaculum Tepidum* [4]) as well as about 100 m under the water surface in the Black Sea at minute light intensities [5]. They have been found even in the deep-sea near hydrothermal vents (black smokers) in depths below 2000 m. This was the first observation that natural photosynthesis is also possible in the complete absence of sunlight, just with the near infrared thermal radiation of up to 370 °C hot springs [6].

Although the internal structure of these antenna systems could not have been completely resolved, recent TEM images from Ganapathy et al. and Oostergetel et al. [3, 7] revealed new insights into the chlorosomes (Fig. 2.1), which led to a new structural model.

The striations all along the chromosomes in the TEM images from Ganapathy et al. have been investigated earlier by Pšenčík et al., who proposed, also based on X-ray measurements, a lamellar model of the BChl *c* self-assemblies [8] with a stack distance of 2.09 nm. The observation that the striations do not disappear upon tilting, together with the results of small angle X-ray scattering (SAXS) measurements, which showed that the long range order is only 6–8 nm, were interpreted, that the lamellas may be bended within the chlorosome.

Ganapathy et al. and Oostergetel et al. [3, 7] could later find new evidence for the earlier proposed tubular model of the antenna structure by new TEM investigations of chlorosome cross-sections. However, the *Chlorobaculum tepidum*

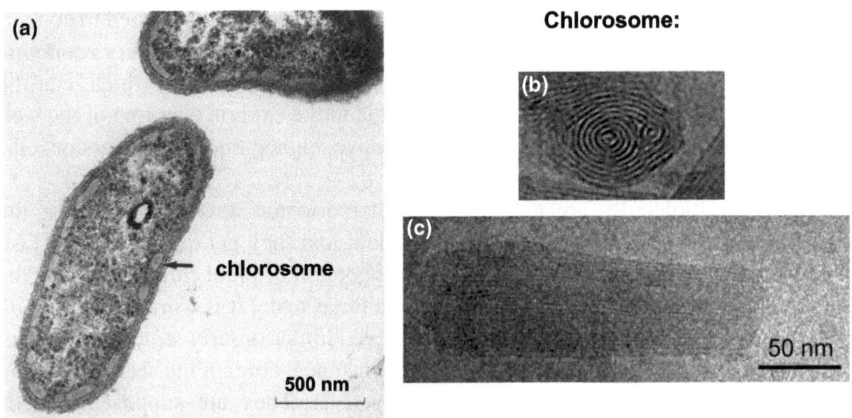

Fig. 2.1 TEM images of two green sulfur bacteria and isolated chlorosomes. **a** TEM image from Wahlund et al. [4] showing two green sulfur bacteria from the species *Chlorobaculum tepidum* (earlier known as *Chlorobium tepidum*). The chlorosomes (one is indicated by the arrow) are located all around the inner side of the cell wall. Copyright (1991) Springer-Verlag. **b** TEM image from Ganapathy et al. [3] of a chlorosome cross-section perpendicular to its long axis bearing a tubular structured antenna system inside. Copyright (2009) National Academy of Sciences, USA. **c** TEM image from Oostergetel et al. [7], showing one chlorosome of the same species in its full length. Striations are visible all along the long axes. Copyright (2007) Federation of European Biochemical Societies

species with the most pronounced tubular shapes were mutants. Both models can be seen in the schematic from Pšenčík et al. and Oostergetel et al. in Fig. 2.2.

Frigaard et al. [9] illustrates in a detailed schematic of the chlorosome in its envelope the earlier proposed model from Rossum et al. and Steensgaard et al. [10, 11] of the BChl *c* assemblies in form of rod-shaped bilayer tubes (Fig. 2.3).

Although much effort was put into the structural analysis of BChl *c* assemblies, its structure could not be completely resolved, due to the lack of long range order. For single-crystal analysis by X-ray diffraction (XRD), these aggregates are too small and electron diffraction is not possible due to an insufficient resistance against beam damage.

2.2 Biomimetic Self-Assembly

Balaban et al. developed the first fully synthetic self-assembling mimics of bacteriochlorophyll *c* [12]. Up to that time only semisynthetic derivatives of chlorophyll *a* could be programed to self-assemble. The full synthetic molecules benefit from relatively inexpensive starting materials available on the market, like the 10,20-bis-(3,5-di-*tert*-butylphenyl) copper porphyrin [12, 13]. They are shown together with their natural counterparts in Fig. 2.4. These molecules were designed to combine the self-assembling characteristics of the BChls with the robustness of artificial zinc porphyrins. The self-assembling bacteriochlorophylls are actually not porphyrins, but so-called chlorins, which are defined to have one of the four pyrrole rings reduced (single bond instead of a double bond between the carbon atoms 17 and 18). Porphyrins instead have a fully conjugated and aromatic π-electron system and thus are very stable. Their stability could be increased by substitution of the central atom

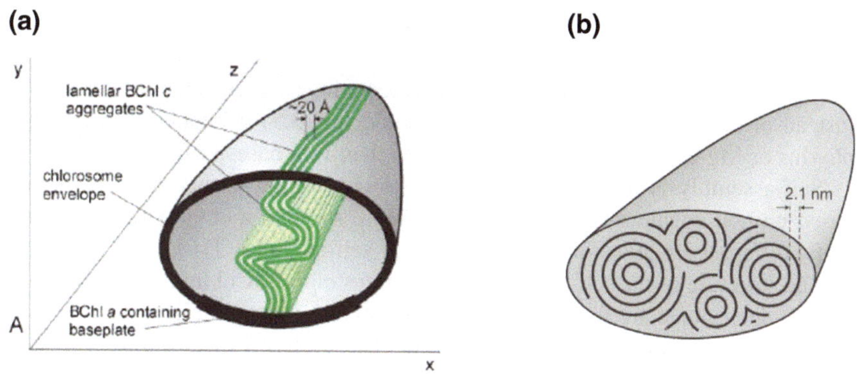

Fig. 2.2 Chlorosomal aggregate structure, **a** Lamellar model of the BChl *c* aggregate structure within a chlorosome envelope after Pšenčík et al. [8], copyright (2004) by the Biophysical Society. **b** Tubular model after Oostergetel et al. [7], copyright (2007) Federation of European Biochemical Societies

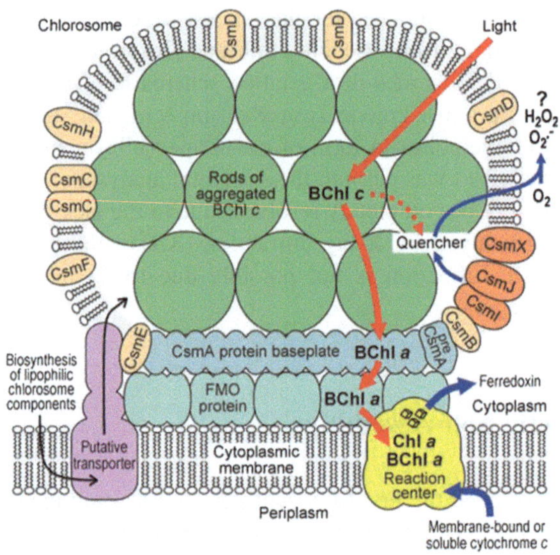

Fig. 2.3 Schematic of a chlorosome cross-section with the tubular BChl *c* aggregate model. This drawing from Frigaard et al. [9] illustrates the components of a chlorosome within its envelope and some of their functions. The tubular model of the aggregate structure of bacteriochlorophyll *c* (BChl *c*) is shown in green, as cross-section of a stack of 10 tubular aggregates. Proteins are embedded within the envelope of the chlorosome. The excitons diffuse through the tubular BChl *c* aggregates toward the BChl *a* within the baseplate. As the absorption spectrum of BChl *a* is red-shifted compared to BChl *c* it acts as a trap for excitons, which then diffuse further to the reaction center (yellow). Oxygen acts as a quencher for the excitons, thus it is shunned by the bacterium. Different concentrations of carotenoids, depending on the bacterium species and the habitat, supplement the light absorption in the green spectral range. Reprinted with permission from Frigaard N-U and Bryant DA (2004) Seeing green bacteria in a new light: genomics-enabled studies of the photosynthetic apparatus in green sulfur bacteria and filamentous anoxygenic phototrophic bacteria, Archives of Microbiology 182:265-276. Reproduced by permission of Springer Science+Business Media. Fig. 2.1

magnesium by zinc, which now resists stronger acids. Also in nature BChls with Zn instead of Mg have been found in very acidic (pH 3.5) habitats, where Mg-porphyrins or Mg-chlorins would decompose to their free bases [14].

Self-assembly is a process in which molecular units form ordered structures based on non-covalent interactions. Argia et al. show in a comprehensive review many different strategies and mechanisms of self-assembly [16].

The structure of the self-assembled aggregates of the BChl mimics in Fig. 2.4 is mainly determined by the type and position of the functional groups, which are attached to the numbered carbon atoms of the porphyrin macrocycle. The same position of the 3-hydroxyethyl group was maintained while the 13^1 keto group in the isocyclic ring was replaced by a 13-acetyl group.

The porphyrin no. 2, called: P2, has been investigated in this thesis toward its suitability for solar cell applications. It forms rod-shaped aggregates as shown in

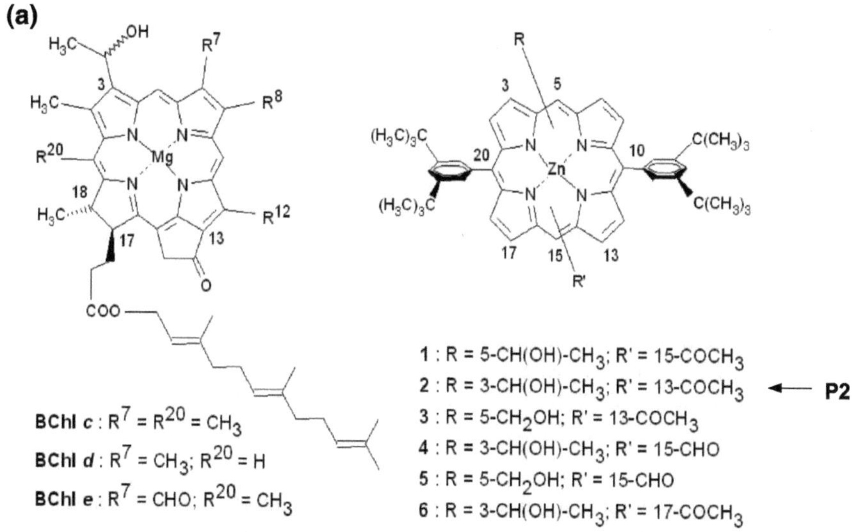

Fig. 2.4 Natural self-assembling bacteriochlorophylls and Balaban's artificial mimics. This graphic from Balaban et al. [15] shows the known self-assembling bacteriochlorophylls (BChl) *c*, *d*, and *e* (left) in comparison to six of his artificial self-assembling mimics (right). The hydrophyilic parts R and R' in the BChl mimics, which are very important for the self-assembly, have been varied in their type (CH(OH)-CH_3 = hydroxyethyl, CH_2OH = hydroxymethyl, $COCH_3$ = acetyl, COH = carbonyl) and their position on the porphyrin macrocycle with commonly numbered carbon atoms from 1 to 20. The biomimetic porphyrin no. 2, here called: P2, has been investigated in this thesis. The central atom magnesium of the BChls was substituted by zinc to increase the stability. The two tert-butylphenyl groups at the carbon atoms 10 and 20 additionally support the self-assembly and increase the solubility in organic solvents. Copyright (2005) Wiley–VCH Verlag GmbH&Co. KGaA, Weinheim

Fig. 2.5a. Porphyrin no. 3, in which, contrary to P2, the hydroxyethyl group at position 3 is substituted by a hydroxymethyl group at position 5, shows a completely different aggregate structure (Fig. 2.5b) [17]: small flakes, which agglomerate to flower-like structures. Also the small change of the hydroxyl group (OH) of P2 into a carbonyl (C = O) one, resulting in a second acetyl group [$COCH_3$] can have a big effect on the self-assembly (Fig. 2.5c). These aggregates of P2-diacetyl have a significantly lower aspect ratio and grow much larger than those of P2, large enough for a single-crystal XRD analysis, which allowed to resolve their structure [18].

The contribution of the hydrophobic groups to the self-assembly was studied by Hasobe et al. by enabling similar porphyrins, but without the oxygen groups and the metal in the center, also to self-assemble by the assistance of ultrasound. Similar results have been obtained also using bare phenyl groups on both opposed sides of the porphyrins (without the *tert*-butyl groups). However, without the treatment in ultrasound, no aggregate formation could be observed [19].

This shows that for self-assembly of the synthetic mimics, and most probably also of the natural BChl models, hydrophilic and hydrophobic, or in general

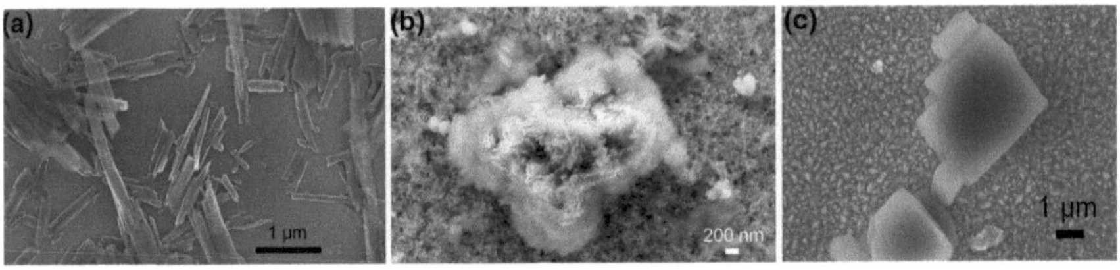

Fig. 2.5 SEM micrographs of P2, P2-diacetyl and Por. No. 3. **a** P2 aggregates deposited by drop-casting from a 2.1 mM solution in anhydrous dichloromethane (DCM) onto a silicon wafer. **b** Aggregates of porphyrin no. 3 from Fig. 2.4 on a mesoporous TiO_2 surface after dip-coating from its anhydrous DCM solution, containing also *n*-heptane, which did not influence the structure. Difference to P2: hydroxyethyl group replaced by hydroxymethyl group and moved from position 3 to 5 in the porphyrin macrocycle. **c** P2-diacetyl aggregates after drop-casting a fresh solution in anhydrous DCM onto a rough SnO_2:F substrate. The size distribution of the cubic-shaped aggregates ranged from ca. 1 µm to ca. 15 µm

solvophilic and solvophobic, interactions play an important role. Whereas the (latent) solubilizing hydrophobic side groups are useful for microcrystallization on the mesoscopic level, they are sometimes detrimental for macroscopic or long-range ordering [20]. However, a long-range order is not always desired as this facilitates the aggregate growth to dimensions, which exceed the exciton diffusion length. On the other hand, when the aggregate's size is too small, this might indicate a highly disturbed internal structure, and inefficient transport of excitons may be the consequence. Thus, an intermediate aggregate size seems to be most promising as it is a compromise between a low defect concentration and a hampered aggregate growth. A tubular structure seems to be a good possibility to combine a low disturbed structure with a limited aggregate size, anyway. When the aggregate grows in a circle, its size will be limited by closure of the circle. The growth within the long axis of the tubes, however, is allowed to exceed the exciton diffusion length as the excitons have a short distance toward the interface in the radial directions, anyway. Thus, tubules may be the ideal architecture for self-assembling antenna systems.

P2 has been chosen for the investigations in this thesis, because it forms elongated aggregates with a high aspect ratio, a high degree of order, which is combined with a shorter pathway for excitons within radial directions (Fig. 2.5a). The same recognition groups (hydroxyethyl and acetyl) were appended in similar positions (3, 13, respectively) as in the natural BChls.

P2 self-assembles to rod-like aggregates of a broad polydisperse distribution in size with a mean length of 1–3 µm and a mean diameter of 100–300 nm (Fig. 2.5a). The mean aggregate size is roughly 10 times larger as the natural antenna system, which is confined within the chlorosomal envelope. Finding ways to confine the aggregate diameter toward the exciton diffusion length, which is supposedly within the range of the chlorosomal thickness, around 20 nm, was also the aim of this work.

2.2.1 Structure of the Self-Assemblies

Similar to the self-assembling bacteriochlorophylls the structure of P2 aggregates could not be resolved completely until now, because the aggregates did not grow large enough for a single crystal analysis, for example, by synchrotron-XRD. P2-diacetyl instead, which forms larger aggregates, as could be seen in Fig. 2.5, did form a crystal sufficiently large for a single crystal structure analysis by synchrotron-XRD as reported by Jochum et al. [18]. With this structure and additional Fourier transform infrared spectroscopy measurements (FTIR) it could be shown, that these molecules form stacks dominated by hydrophobic interactions (Fig. 2.6). The electrostatic interaction between the carbonyl groups and the zinc atoms is rather small as the acetyl groups do not turn significantly out of the conjugation plane, which they have in common with the porphyrin macrocycle.

By comparison of the coupling strength between the carbonyl groups and the zinc atoms, measured by FTIR in various porphyrins, including P2 and self-assembling BChls, it could be concluded that contrary to previous assumptions no hydrogen bonds exist between the porphyrin stacks [18]. The old and the new model of the basic aggregate structure of self-assembling BChls and their artificial mimics is shown in Fig. 2.7.

In the new structural model for BChl self-assemblies (Fig. 2.7b) the Mg atoms in the porphyrin centers are assumed to be strongly coordinated by a hydroxyl group from one neighboring molecule and weakly coordinated by a carbonyl group from another neighbor in the stack resulting in a 5½ coordination. This new structure without hydrogen bonds was also assumed for the P2 aggregates. It allows also for a high mechanical flexibility [21], which has often been observed for the P2 aggregates.

Due to the encountered electron beam damage this structural model for the BChls could not be verified by electron diffraction, yet.

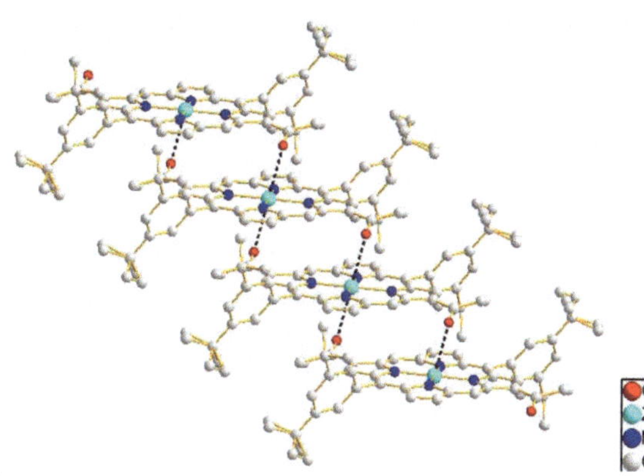

Fig. 2.6 Model of the P2-diacetyl crystal structure. This model structure of P2-diacetyl-aggregates from Jochum et al. [18] has been determined by single crystal XRD. Copyright (2008) National Academy of Sciences, U.S.A

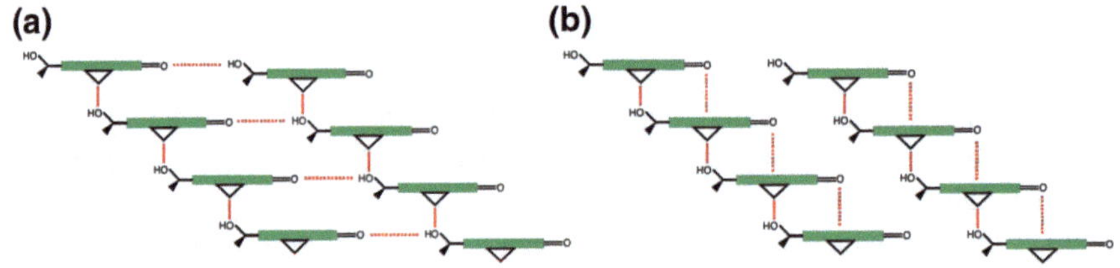

Fig. 2.7 Old and new model of BChl self-assemblies. These schemes from Jochum et al. [18] show the previously assumed structure of bacteriochlorophyll (BChl) self-assemblies **a** and the new proposed structure model **b**, which is also representative for the artificial mimics, like P2 aggregates. This structure was proposed by these authors as a conclusion from the resolved aggregate structure of P2-diacetyl in combination with additional FTIR studies. Copyright (2008) National Academy of Sciences, USA

2.3 Main Characterization Methods

This chapter gives a short introduction into the main characterization methods, UV–Vis–NIR absorption spectroscopy and scanning electron microscopy (SEM).

2.3.1 Absorption Spectroscopy

The absorption was investigated within the wavelength range from the ultraviolet over the visible up to the near infrared part of the electromagnetic spectrum. The absorption spectrum at the longest wavelength of organic chromophores or chromophore aggregates originates by the electronic transition between the highest occupied molecular orbital (HOMO) and the lowest unoccupied molecular orbital (LUMO). The whole spectrum is a superposition of electron excitations also from ground energy levels beneath the HOMO (HOMO-1, -2,…) and to excited energy levels above the LUMO (LUMO+1, +2,…) or into higher vibronic levels of the electronic states. Due to additional vibronic modes within liquids and solids the absorbance maxima are broadened, compared to molecules in the gas phase.

The absorption spectrum is commonly given as absorbance over wavelength. Whereas the direct unit of measure is the transmittance, which is the ratio of incident to transmitted spectral radiant power (intensity). The relation between absorbance, transmittance and the quantity of the absorber is given by the Beer–Lambert law:

Beer–Lambert law: $A_{10} = c\,l = -\log_{10}(T)$with: $T = \frac{I}{I_0}$ or in the form: $I = I_0\,10^{-cl}$ where A_{10} is the decadic absorbance, ε the decadic molar absorption coefficient, c the molar concentration, l the path length of the cuvette measured in cm, I the transmitted and I_0 the incident spectral radiant power, and the ratio is the transmittance T.

In the case of solid samples (instead of liquids) $\varepsilon \cdot c$ is replaced by the absorption coefficient α. l does then mean the thickness of the absorbing layer instead of the path-length of the cuvette. Although the natural logarithmic form of the Beer–Lambert law is commonly used for solid samples, throughout this work only the decadic form was used in order to avoid confusion.

The term absorbance (A) is only valid, when no "light" scattering or luminescence contributed to the absorption spectrum. Otherwise this term is to be replaced by the attenuance (D) and ε (α) will be called (molar) attenuation coefficient in this case [22].

The following devices were used for the absorption spectroscopy:

UV–Vis–NIR absorption spectrometer: *cary 500 Scan* from *Varian*

Integrating sphere: *DRA-CA-5500* with 15 cm Ø

2.3.1.1 Light Scattering Samples

Absorption spectroscopy of light scattering samples, like turbid dispersions and rough or porous solids, is in general more difficult to interpret as non-scattering solid or liquid samples, even when measured within an integrating sphere.

The integrating sphere can be described as a collector of the scattered light, which otherwise would escape the detector and thus would be interpreted as absorbed fraction. Measurements on scattering samples were performed within the center of an integrating sphere to collect the transmitted and reflected part of the incident monochromatic light simultaneously. The detector, attached to the integrating sphere, measured that intensity after multiple scattering within the sphere.

In contrast to non-scattering samples multiple reflections occur also within the sample causing an uncertainty in the effective path-length for the incident beam. Thus, a quantification of the absorber is not possible. For this purpose, one needs to dissolve the absorber completely and perform a measurement in a non-scattering environment. Alternatively, one could embed or disperse the scattering sample, for example, particles, within a transparent matrix with the same refractive index to eliminate the scattering.

Additionally, large dye aggregates on a substrate or within dispersions can show lower absorbance as smaller ones with the same molar concentration, because some inner dye molecules within larger aggregates may not be reached by the incident light. Such inner dye molecules do not participate in the absorption process and simulate a lower absorption coefficient for the whole system.

The morphology of the system does not only affect the quantitative absorption, but also the qualitative absorption spectrum, in the case of aggregates, for example, this is caused by exciton coupling of adjacent chromophores and depends on the mutual orientation. Therefore, amorphous and crystalline aggregates show different absorption spectra.

2.3.1.2 Porphyrins

The absorption spectra of zinc porphyrin monomers show a characteristic sharp and intense band in the blue part of the spectrum, the so-called Soret band, and typically two smaller bands in the red part, called Q bands (Fig. 2.8a). The origin of the corresponding transitions is described by the so-called four-orbital model of Gouterman [23]. These orbitals are the two HOMOs, called b1 and b2 and the two LUMOs c1 and c2. Due to the lower energy level of b1, which is a π-orbital of the symmetry a_{2u} (π), the corresponding energy level is assigned to HOMO-1 and the π-orbital b2 with the symmetry a_{1u} (π), is the HOMO [24, 25]. In metalloporphyrins c1 and c2, which both have the symmetry $e_{.g.}$ (π^*) are degenerate, that is, of the same energy. The degeneracy arises from the equivalence of the classical dipoles in the perpendicular directions x and y within the D_{4h} symmetrical porphyrin plane [23]. The Soret (B) band transition occurs from the HOMO-1 to the LUMO, whereas the Q band originates from the transition between the HOMO and the LUMO. The corresponding two pairs of degenerate transitions with perpendicular polarization are called (B_x, B_y) and (Q_x, Q_y). Their transition dipole moments are polarized (oriented) within the porphyrin plane like shown in Fig. 2.8b, where x' and y' indicate a second coordinate system.

Besides the excitation between the lowest vibrational levels of the two electronic states, denoted by (0,0) [26] also excitations into higher vibrational states of the LUMO can occur. As the Q_x transition couples strongly vibronically with the Soret band transition it has a higher intensity for the transition into the first excited vibrational state of the LUMO, than Q_y. It is therefore denoted also by Q (1,0). This gain in intensity is also called vibronic borrowing [27, 28]. Q_y instead, does not borrow much intensity from the Soret band and corresponds predominantly to

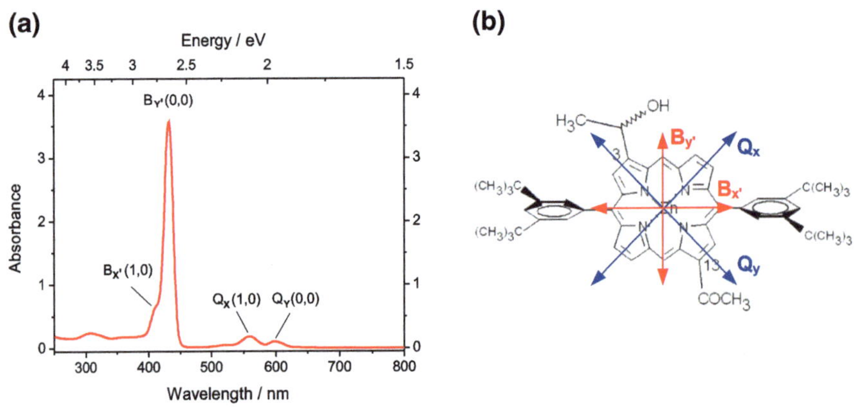

Fig. 2.8 Absorption spectrum of P2 monomers. **a** This absorption spectrum of the zinc porphyrin P2, dissolved in tetrahydrofuran, represents a typical absorption spectrum of metalloporphyrins. Whereas the numbers in brackets denote the vibrational states of the LUMO and the HOMO, respectively, the relative orientations (polarizations) of the corresponding transition dipole moments are shown in. **b** with x' and y' indicating a second coordinate system

the transition into the vibrational ground state of the LUMO and is therefore denoted also by Q (0,0). It might be confusing that different notations are used in literature to denote the same transition, for example, for Q (1,0) [28, 29] also the following notations can be found: Q (0,1) [30], Q (0-1) [23] or Q (1-0) [31]. Figure 2.8a shows a typical absorption spectrum of metalloporphyrins.

2.3.1.3 Exciton Coupling Theory

The absorption spectrum of aggregates differs from that of monomers, due to the exciton coupling between neighboring chromophores, depending on their orientation next to each other. This leads to a shift of the absorption spectrum, compared to that of isolated chromophores, either toward higher energies (blue-shift), or toward lower energies (red-shift). The corresponding aggregate types are called H-aggregates, and J-aggregates, respectively. H-aggregates are named after the term: hypsochromic, which means: shifted toward higher energies, and J-aggregates after E.E. Jelly, one of the first workers, who investigated the red-shift, which is also called bathochromic shift. Both shifts may occur also simultaneously, as the molecules couple to several others within three dimensional aggregates. As amorphous aggregates do not have a defined structure their absorption spectrum does not show any characteristic shift, but is simply broadened. Absorption spectra of isolated monomers within perfect solutions may also be shifted slightly depending on the polarity of the solvent molecules. Thereby the whole spectrum is shifted in contrast to the case of well defined aggregates where the shift is band-specific [32].

The shift of the absorption spectrum results from the splitting of the excited state (LUMO), caused by Coulomb interactions of adjacent molecules, and from the selection rule of the transition dipole moments, like shown in the illustration from Satake et al. in Fig. 2.9.

The transition dipole moment (arrow) within a planar dye molecule, like a porphyrin, interacts with the transition dipole moment of an adjacent molecule due to Coulomb interactions, leading to a splitting of the excited energy state E (LUMO) of a monomer into a higher (E'') and a lower (E') energetic state of the dimer. In the case of a pure face-to-face orientation of both molecules the parallel orientation of both transition dipole moments result in a higher energy and the antiparallel dipole moment orientation in a lower one. This can be explained by an electrostatic repulsion of the transition dipole moments in the former and attraction in the latter case [34]. For the head-to-tail orientation of the dye molecules their transitions dipole moments interact in the opposite way because here the parallel orientation of the transition dipoles interact attractively and the antiparallel orientation repulsively. As the transition to an energy state with antiparallel transition dipole moments is not allowed, as they would cancel out each other, only the transition to the energy level with parallel oriented transition dipole moments will occur, what leads to the blue- or red-shift in the absorption spectrum depending on the molecular orientation. The slipped cofacial arrangement is a mixture of both,

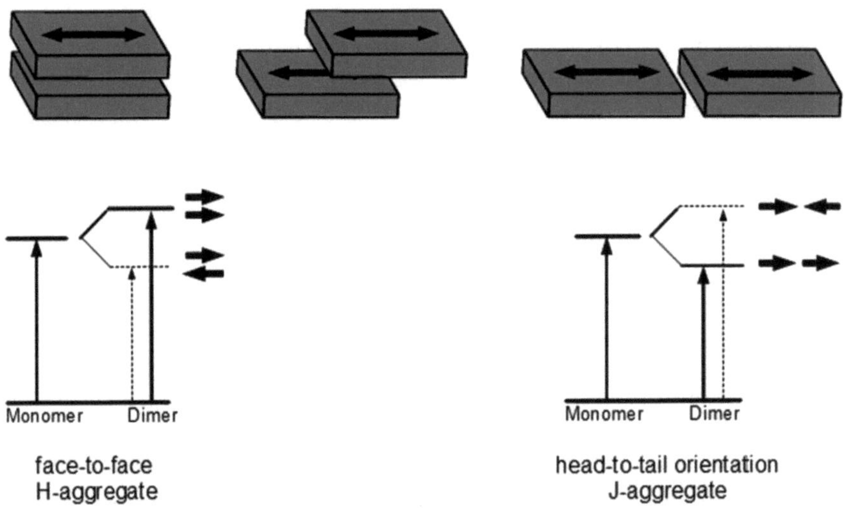

Fig. 2.9 Illustrated explanation of the *red*- and *blue*-shift of aggregates. These cartoons illustrate how the arrangement of two planar identical molecules (e.g., porphyrins) to a dimer and thus the arrangement of their dipole moments (arrows) influences the excited state energy splitting and selection rules, which correspond to a *red*- or *blue*-shifted absorption spectrum. The thick horizontal arrows indicate a momentum view of the transition dipole moments. The thin solid vertical arrows indicate the allowed transitions from the ground states (HOMO) to the excited states, whereas broken lines correspond to forbidden transitions. The energy splitting (distance between both excited states of a dimer) is dependent to the angle between the molecules and can reach zero, as shown in Fig. 2.10. [33]

which may cancel out the energy splitting. In principle a certain orientation or stacking may be extended to many more molecules, forming so-called H- or J-aggregates, assigning the blue- or red-shift in the absorption spectrum, respectively.

The pure face-to-face and head-to-tail orientations are two extrema in the stacking possibility of the molecules with the maximum splitting energy ΔE between *E'* and *E"*. In intermediate positions, by cofacially slipping of the molecules from one extrema to the other, *E'* continuously goes over to *E"* and vice versa, through an orientation with no energy splitting. ΔE can be calculated by the following formula [33]:

$$\Delta E = \frac{2|M_1 M_2|\mathrm{k}}{r^3} \text{ with } \mathrm{k}^2 = (1 - 3\cos^2\theta)^2$$

where M_1 *and* M_2 are the transition dipole moments of both molecules, which are the same in the case of identical molecules, r *is* the center-to-center distance between both molecules, κ the orientation factor and θ is the angle between both molecules (transition dipole moments) like shown in Fig. 2.10.

The curve in Fig. 2.10b shows the orientation factor κ over the angle θ, which indicates the arrangement of both molecules in the dimer, starting from the face-

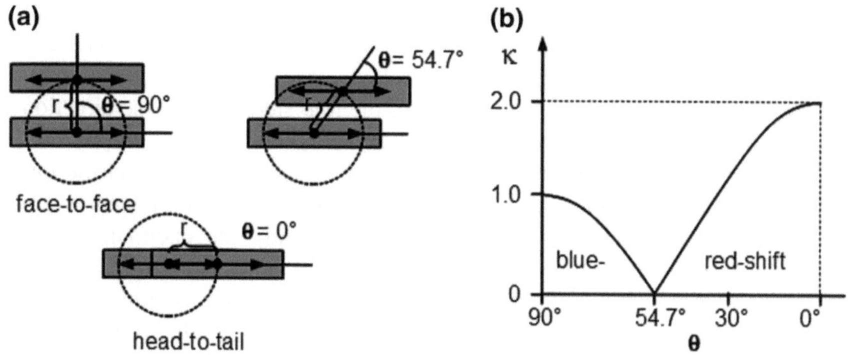

Fig. 2.10 Dependency of the energy state splitting on the slipped cofacial molecular arrangement. This figure illustrates the dependence of the excited state splitting ΔE, expressed by the orientation factor κ, **b**, on the change in the slipped cofacial arrangement, expressed by the angle Θ **a**. The arrows represent the orientation of the transition dipole moments of two molecules in a dimer with r as the center-to-center distance between them. While the molecules slip cofacially from the face-to-face orientation into the heat-to-tail one, the *blue*-shift goes over into a *red*-shift at $\Theta = 54.7°$, the point at which the excitation energy state is not split. [33]

to-face arrangement at $\theta = 90°$ and ending at the head-to-tail arrangement at $\theta = 0°$. With slipping the molecules from the face-to-face arrangement into the head-to-tail one, the blue-shift continuously decreases and turns at $\theta = 54.7°$ into the red-shift, which increases steadily to reach the maximal red-shift at the pure head-to-tail arrangement.

The shift of the absorption bands is not necessarily an indication for a change in the arrangement of the dye molecules. In small aggregates beside the band intensity also its shift is dependent to the number of the chromophores. Osuka, Kim and their coworkers [28, 35] on covalently bound porphyrins, arranged in chains with the pure head-to-tail orientation, the dependence between the energy splitting ΔE_0 and the number N of chromophores to:

$$\Delta E_0 = \cos[\pi/(N + 1)]$$

Hence, the red-shift of pure head-to-tail arranged porphyrin chains starts with a dimer already at 50 % of the maximum shift, increasing fast at the beginning reaching 90 % in a chain of six chromophores and with 99 % almost the maximum at 22 chromophores.

In contrast to non-covalently bound aggregates, these structures are possible to be obtained monodisperse as they can be separated, for example, by chromatography. Porphyrin aggregates, instead are relatively weakly bound molecules, which are in a dynamic equilibrium with the solvent, thus can not be separated to a monodisperse size distribution. Additionally, the arrangement within chains is only one dimensional avoiding any exciton coupling within other dimensions, which is the case in three dimensional aggregates. Therefore, these model structures of Aratani et al. are well suited for investigations of the exciton coupling effects [35, 36]. In Ref. [36] Aratani et al. show schematically the arrangement of the linked zinc porphyrins and the resulting band splitting diagram. These porphyrins are covalently bound in the so-called meso position of the porphyrin macrocycle, which is in the center of each of the four sides, thus in the direction of the transition dipole moment of the B (Soret) band, here: B_x. The porphyrins are additionally rotated by 90° around this axis, which excludes any contribution of the otherwise blue-shifting face-to-face orientation of the B_y component.

As this one-dimensional model system allows only the excitonic coupling within this one dimension (x) only the B_x part of the Soret band shifts to the red part of the spectrum, while for B_y and B_z the monomeric position of the LUMO does not split. The corresponding absorption spectroscopy results from Kim et al. show also the dependency of the red-shift on higher numbers of meso-linked porphyrins Fig. 2.11.

Figure 2.11 shows that the linkage of two molecules already leads to a significant red-shift and broadening of the absorption spectrum, what is very favorable for solar cell applications. This experiment tells us also that the desired red-shift and broadening is almost complete after coupling of a few molecules, as can be seen in the here found correlation factor: $\cos[\pi/(N+1)]$. The broadening of the B band indicates a heterogeneous distribution of the dihedral angle at the linkage between the porphyrins, which allows also a partly excitonic coupling of the B_y

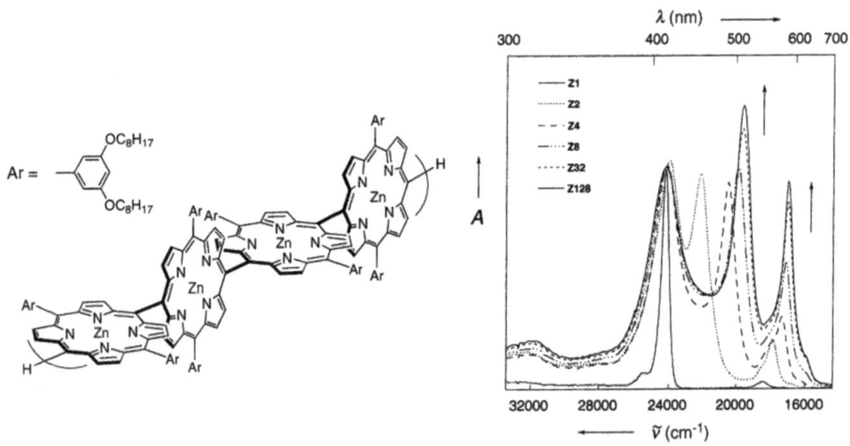

Fig. 2.11 Absorption spectra of meso–meso-linked zinc porphyrin chains. Aratani et al. [28] show in these graphs the dependency of the *red*-shift of the absorption spectrum on the number Z(N) of covalently bound zinc porphyrins. The porphyrins are linked in the so-called meso-position of the porphyrin macrocycle and orient orthogonally to each other. By increasing the number of linked porphyrins from 1 to 128 within the chains, predominantly one part of the Soret band shifts to the *red*-part of the spectrum. The Q band also makes a minor *red*-shift, but with a considerable increase in intensity. Note also the increase in the ultraviolet region and the broadening of the *B* bands. Copyright (2002) Japanese Photochemistry Association

(B_z) component. The angle of 90° was predicted by semi-empirical Austin Model 1 (AM1) calculations as the most stable conformation and was confirmed by X-ray crystal structure analysis. [35]

The dominant Q (1,0) band, originally around 18500 cm^{-1}, only slightly shifted bathochromically but therefore its intensity increased significantly with increase of the chain length. This was attributed by Kim et al. to an increased vibronic coupling with the B band due to the lower energy distance to the red-shifted B band [35]. The red-shift of the Q bands can be explained by the fact, that their transition dipole moments have also a component within the X-direction because they are tilted only by 45° relative to the X-axis. Hence both Q bands ($Q_{x'}$ and $Q_{y'}$) couple partly in the head-to-tail way, splitting their LUMO and leading to a partly red-shift. This agrees with the fluorescence anisotropy measurements of Kim et al., which led to the conclusion that the polarization of the red-shifted Q bands is the same as that of the red-shifted B band [35].

This example showed that exciton coupling in zinc porphyrins, even with a low number of involved chromophores, leads to a significant red-shift and broadening of the absorption spectrum. This may be regarded as a bonus for solar cells as they would already benefit from the exciton transport property of such systems due to the possibility to reduce the crucial interface area in dye sensitized solar cells (DSSCs). Thus, these so-called supramolecules have very favorable optoelectronic properties for solar cells. However, their aggregation may be problematic, as large molecules tend to aggregate to amorphous solids. It is important, that the well-defined structure within the supramolecules is not disturbed by the formation of disordered aggregates, which would increase the exciton quenching rate at the interfaces between these supramolecules. Additionally, the synthesis of such supramolecules is difficult, as they tend to loose solubility with increasing size. Also low yields, difficulties in separation and demanding characterizations hamper the progress exploring such molecules on larger scales [28].

Self-assembling chromophores, however, have the advantage of an easier synthesis, a higher solubility and a controlled self-assembly to well defined aggregates preserving their functionality also in the solid-state. Additionally, self-assembling chromophores may be infiltrated more easily into porous structures, like in DSSCs, compared to long chains of covalently bound chromophores.

It can be concluded that beside the well understood red-shift of parts of the absorption spectrum, some effects still remain partly unpredictable, like the band broadening and changes in their intensities. Nevertheless, the covalently bound macromolecules are a very good model system, due to their one-dimensional exciton coupling in the pure head-to-tail fashion and a precisely controlled number of chromophores. Compared to this relatively simple model it can be imagined, that it the exciton coupling effects on the absorption spectrum become far more complicated to analyze in the case of three dimensional aggregates.

2.3.1.4 Scanning Electron Microscopy

This chapter gives a short introduction into the SEM technique and an overview about the typical morphology of P2 aggregates. Beside the optical indications that P2 self-assembles to an ordered J-aggregate structure also the morphological investigations by the SEM could affirm the well-defined structures by visualizing their regular shapes. As the SEM was a major tool for investigation of the P2 J-aggregates, this technique will be described in more detail.

The used scanning electron microscope was a *LEO 1530 Gemini* from *Zeiss*. It was equipped with two detectors for secondary electrons (SE), the so called in-lens detector, which acquires only slow secondary electrons and the SE2 detector, also called lateral detector or Everhart–Thornley (ET) detector [37], for faster ones, which escape also from regions deeper under the surface.

This SEM allows to focus the primary electron beam onto a spot with the diameter of around 2 nm, depending, on the aperture size and the accelerating voltage. The primary electrons (PE) penetrate the sample to a depth, up to a few μm, dependent mainly to the accelerating voltage of the electron beam, the density of the sample and the atomic numbers of its composition. Many secondary electrons (SE) can be hit out of the sample's atoms by one PE impact, but only those SEs, which got hit out close to the surface, can escape to the vacuum and be detected. As most SEs have kinetic energies below 50 eV their mean escape depth is in the order of a few nanometers. Compared to metals, materials of lower electron density, like ceramics or organics have a larger escape depth, which depends also on the material's structure. Thus, P2 aggregates with a highly ordered internal structure appear more transparent for SEs than amorphous ones, especially when looking for low-energy SEs, which are detected by the in-lens detector.

Three different types of secondary electrons (SE1-3) are distinguished [38, 39]:

- SE1 electrons originate by collisions with primary electrons (PE). Their detection generates the highest lateral resolution.
- SE2 electrons originate by collisions with backscattered electrons (BSE). Their topographical information is of a lower resolution, but their contrast generation offers additional information.
- SE3 electrons originate by collisions of the BSEs with the gun or the chamber wall. They generate the background noise.

The detected signal is always a mixture of different secondary electron types, mainly SE1 and SE2. The in-lens detector allows for a higher resolution then the SE2 detector, because it gathers a higher fraction of the SE1 electrons compared to the SE2 detector. As it is located within the final lens of the electron gun, it is much closer to the specimen compared to the SE2 detector. Due to the resulting increase of the signlal-to-noise ratio it is possible to lower the accelerating voltage of the PE, what increases the ratio of SE1 to SE2 [37]. As the in-lens detector gathers those filtered SEs with lower energy it is more surface sensitive than the SE2 detector. The SE2 detector attracts with higher positive potential (some hundred eV) also faster SEs with a higher escape depth and offers therefore a more semitransparent appearance of the specimen. Additionally, it generates, due to its lateral location at one side of the chamber, also a shadow contrast, which gives the sample a more three dimensional impression.

Beside the detectors SE2 and in-lens the following parameters are used to adjust the desired ratio of different contrast types, what affect also the resolution:

- accelerating voltage for the electrons, also called extra-high tension (EHT)
- working distance (WD), which is the distance between the final condenser lens and the specimen
- aperture size (aperture), controls the beam current.

A high EHT normally decreases the spherical aberration and thus the resolution, but it increases also the penetration depth and this may also decrease again the resolution. A smaller working distance reduces the spot size due to minimizing the spherical aberration and thus increases the resolution, but this decreases also the field of depth, due to the bigger focusing angle. A smaller aperture size reduces the beam current and thus also the spot size, what increases the resolution, but makes the image more noisy.

A comparison of the material contrast generation by the two detectors is shown in Fig. 2.12, where the same area has been recorded with both, the in-lens (a) and the SE2 (b) detector. The sample was a 0.1 mm thick quartz substrate with 110 nm thick and and 1.8 μm wide vertically aligned gold electrodes in a distance of also 1.8 μm. On this substrate a drop of an anhydrous P2-dichloromethane solution was dried under an applied electrical field. This technique induced a partial separation of the aggregates by size, what will be described in more detail in the Chap. 4.9. (page 83). The P2 aggregates appear dark in the micrograph of the in-lens detector. In the SE2 detector micrograph, instead the small aggregates are difficult to

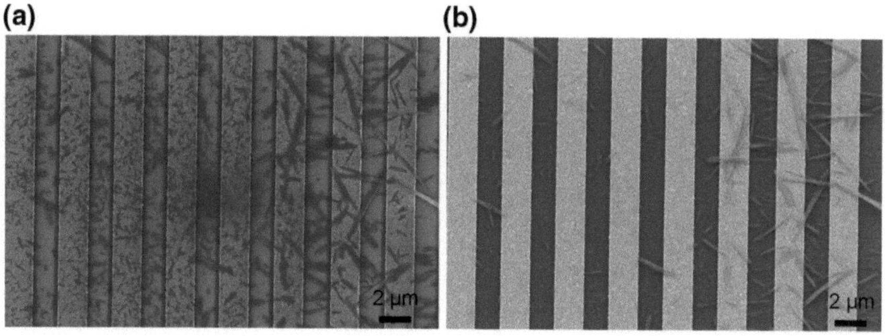

Fig. 2.12 SEM micrographs showing the contrast difference of the in-lens and SE2 detector. Both micrographs show exact the same area of P2 aggregates on a 0.1 mm thick quartz substrate with vertical stripes of a 110 nm thick gold layer, but recorded with different detectors, the in-lens **a** and the SE2 detector **b**. EHT: 5 kV, WD: 5 mm, aperture: 30 μm. The gold structure was prepared by standard photolithography and the aggregates were deposited by dielectrophoretic controlled drop casting of a P2- dichloromethane solution

recognize, but therefore the material contrast between gold and quartz is more pronounced. The bigger aggregates instead, appear with a higher plasticity and three dimensional impression in the SE2 micrograph.

The dark spot in the middle of the in-lens micrograph (Fig. 2.12a) is caused by the deposition of a thin organic contamination layer, which appeared during focusing in this area, prior to the image recording. This effect occurs when volatile organic molecules in the gas phase of the limited vacuum polymerize onto the sample's surface after being radicalized by the electron beam. The strong contrast between the thin contamination layer and the rest of the surface demonstrates the high sensitivity of the in-lens detection for such thin organic layers, which are supposed to be not thicker, than a few nm. Note that the dark spot is not noticeable in the SE2 micrograph (**b**) although it has been recorded after the in-lens micrograph. This high surface sensitivity is not possible with the SE2 detector, which records faster secondary electrons, which originate from deeper regions, enhancing the material contrast of thicker layers, like between the gold stripes (bright) and the quartz substrate (dark). Note also the higher plasticity in the SE2 micrograph, giving a better three dimensional impression, whereas the in-lens micrograph gives a higher lateral resolution and material contrast of thin films. Both detector signals can also be mixed during the image recording to benefit from the advantages of both detectors within one micrograph.

Although the exposed quartz substrate between the grounded gold electrodes is highly insulating the distance to the electrodes was small enough to discharge the quartz surface during the SEM investigation. Hence, no charging effects could be observed here, in contrast to quartz areas, which were a few μm further apart from any grounded electrodes.

The depth sensitivity of both detector signals is demonstrated in Fig. 2.13, where a mesoporous, several μm thick, TiO_2 layer, equipped with additional homogeneously dispersed macropores of 1 μm diameter, was imaged by the in-lens (a) and the SE2 detector (b) at an acceleration voltage of 20 kV. Whereas the

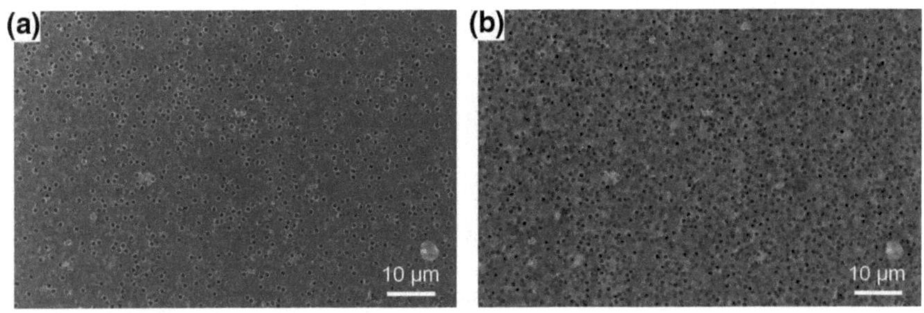

Fig. 2.13 SEM micrographs showing the depth effect of both detectors (in-lens, SE2). The same area of a mesoporous TiO_2 layer with homogeneously distributed 1-μm-sized macropores imaged by the in-lens (**a**) and the SE2 detector (**b**). The SE2 detector reveals also the presence and distribution of inner macropores beneath the surface. EHT: 20 kV, WD: 11 mm, aperture: 30 μm

in-lens micrograph shows only the open macropores at the surface, in the SE2 micrograph also the inner macropores beneath the surface appear as dark spots.

Such meso- and macroporous structures have been used to confine the P2 aggregate size, which will be shown in more detail in Chap. 4.7.2.

References

1. Balaban TS (2005) Acc Chem Res 38:612–623
2. Blankenship RE (2002) Molecular mechanisms of photosynthesis. Blackwell Science Ltd, Oxford
3. Ganapathy S, Oostergetel GT, Wawrzyniak PK, Reus M, Chew AGM, Buda F, Boekema EJ, Bryant DA, Holzwarth AR, De Groot HJM (2009) Proc Natl Acad Sci USA 106:8525–8530
4. Wahlund TM, Woese CR, Castenholz RW, Madigan MT (1991) Arch Microbiol 156:81–90
5. Manske AK, Glaeser J, Kuypers MMM, Overmann J (2005) Appl Environ Microbiol 71:8049–8060
6. Beatty JT, Overmann J, Lince MT, Manske AK, Lang AS, Blankenship RE, van Dover CL, Martinson TA, Plumley FG (2005) Proc Natl Acad Sci USA 102:9306–9310
7. Oostergetel GT, Reus M, Chew AGM, Bryant DA, Boekema EJ, Holzwarth AR (2007) FEBS Lett 581:5435–5439
8. Psencík J, Ikonen TP, Laurinmäki P, Merckel MC, Butcher SJ, Serimaa RE, Tuma R (2004) Biophys J 87:1165–1172
9. Frigaard N-U, Bryant DA (2004) Arch Microbiol 182:265–276
10. van Rossum B-J, Steensgaard DB, Mulder FM, Boender GJ, Schaffner K, Holzwarth AR (2001) Biochemistry 40:1587–1595
11. Steensgaard DB, Wackerbarth H, Hildebrandt P, Holzwarth AR (2000) J Phys Chem B 104:10379–10386
12. Balaban TS, Bhise AD, Fischer M, Linke-Schaetzel M, Roussel C, Vanthuyne N (2003) Angew Chem Int Ed Engl 42:2139–2144
13. Bhise AD (2005) A biomimetic approach for synthesizing artificial light-harvesting systems using self-assembly. PhD thesis, University of Karlsruhe. Forschungszentrum Karlsruhe GmbH, Karlsruhe, http://bibliothek.fzk.de/zb/berichte/FZKA7174.pdf
14. Wakao N, Yokoi N, Isoyama N, Hiraishi A, Shimada K, Kobayashi M, Kise H, Iwaki M, Itoh S, Takaichi S (1996) Plant Cell Physiol 37:889–893
15. Balaban TS, Linke-Schaetzel M, Bhise AD, Vanthuyne N, Roussel C, Anson CE, Buth G, Eichhöfer A, Foster K, Garab G, Gliemann H, Goddard R, Javorfi T, Powell AK, Rösner H, Schimmel T (2005) Chem Eur J 11:2267–2275
16. Ariga K, Hill JP, Lee MV, Vinu A, Charvet R, Acharya S (2008) Sci Technol Adv Mater 9:014109
17. Marek P (2005) Selbstanordnende Zn-Porphyrine in makroporösen TiO_2-Schichten für farbstoffsensibilisierte Solarzellen. Diploma thesis, Technical University of Darmstadt
18. Jochum T, Reddy CM, Eichhöfer A, Buth G, Szmytkowski J, Kalt H, Moss D, Balaban TS (2008) Proc Natl Acad Sci USA 105:12736–12741. http://pubs.rsc.org/en/content/articlelanding/2008/cc/b713971c
19. Hasobe T, Oki H, Sandanayaka ASD, Murata H (2008) Chem Commun 724–726
20. Hoeben FJM, Jonkheijm P, Meijer EW, Schenning APHJ (2005) Chem Rev 105:1491–1546
21. Reddy CM, Padmanabhan KA, Desiraju GR (2006) Cryst Growth Des 6:2720–2731
22. Absorbance, IUPAC, Gold Book, 09/2009, http://goldbook.iupac.org/A00028.html
23. Gouterman M (1959) J Chem Phys 30:1139–1161
24. Gouterman M (1961) J Mol Spectrosc 6:138–163

25. Steiner E, Fowler PW (2006) Chlorophylls and bacteriochlorophylls. Springer, The Netherlands
26. Zero-zero (0-0) absorption or emission, IUPAC, Gold Book, 09/2009, http://goldbook.iupac.org/Z06744.html.
27. Kano H, Saito T, Kobayashi T (2001) J Phys Chem B 105:413–419
28. Aratani N, Osuka A, Cho S-H, Kim D (2002) J Photochem Photobiol, C 3:25–52
29. Spellane PJ, Gouterman M, Antipas A, Kim S, Liu YC (1980) Inorg Chem 19:386–391
30. Ceulemans A, Oldenhof W, Gorller-Walrand C, Vanquickenborne LG (1986) J Am Chem Soc 108:1155–1163
31. Bozkurt SS, Merdivan M, Ayata S (2010) Spectrochim Acta Part A Mol Biomol Spectrosc 75:574–577
32. Dahal S, Krishnan V (1995) J Photochem Photobiol A Chem 89:105–112
33. Satake A, Kobuke Y (2007) Org Biomol Chem 5:1679–1691
34. Buwalda RT (2001) Molecular aggregation in water: the interplay of hydrophobic and electrostatic interactions. PhD thesis, University of Groningen, The Netherlands
35. Kim YH, Jeong DH, Kim D, Jeoung SC, Cho HS, Kim SK, Aratani N, Osuka A (2001) J Am Chem Soc 123:76–86
36. Aratani N, Kim D, Osuka A (2009) Acc Chem Res 42:1922–1934
37. Cazaux J (2003) J Microsc 214:341–347
38. Reimer L, Pfefferkorn G (1973) Rasterelektronenmikroskopie. Springer, Berlin, Heidelberg, New York
39. Seiler H (1983) J Appl Phys 54:R1–R18

Chapter 3
Controlling the Self-Assembly of Zinc Porphyrin P2

The investigations on the self-assembling P2 are described in this chapter. Different experimental methods were used to determine its optical and electronic properties, which depend on the aggregate morphology, such as the aggregate structure and size. Absorption spectroscopy in combination with scanning electron microscopy (SEM) was used in order to determine if there is a correlation between the optical and morphological properties of the aggregates. Such a correlation would be useful later on to quickly infer the average or dominating structure of the aggregates on a larger area and of hidden or embedded aggregates, which are not accessible by the SEM technique. Other properties, which are relevant for solar cell applications, like the aggregate's thermal stability, photoconductivity as well as the possibility to control the aggregate's morphology by different deposition techniques, like spin-coating, spraying and dielectrophoresis were also investigated.

3.1 Structure of the Molecule

The size and structure of the zinc porphyrin P2, has been computed by the semiempirical parametric method 3 (PM3) with the *Hyperchem* program package, version 7.0 HyperCube, Gainesville, Florida (Fig. 3.1). The full chemical name of P2 is: 3-(1-hydroxyethyl)-13-acetyl-10,20-bis-(3,5-di-*tert*-butylphenyl) porphyrinato zinc.

The numbers, which are not in brackets, denote the carbon atoms of the porphyrin macrocycle, conventionally numbered clockwise from 1 to 20. The oxygen containing functional groups for self-assembly are bound at the atoms number 3 and 13. The two 3,5-*tert*-butylphenyl groups, bound to the atoms 10 and 20 of the porphyrin macrocycle, increase the solubility and probably also support the self-assembly.

The porphyrin macrocycle of P2 is planar with the dimensions of approx. 0.7×0.7 nm (between opposing carbon atoms). The distance between the first carbon atoms of the oxygen groups is approx. 1.0 nm, and between the farthest

P. L. Marek, *Biomimetic Dye Aggregate Solar Cells*,
Springer Theses, DOI: 10.1007/978-3-319-00636-9_3,

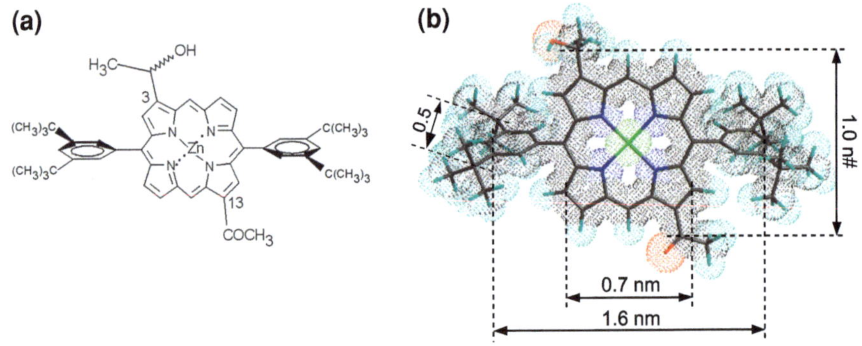

Fig. 3.1 Molecular structure of P2. The chemical structure of P2 (**a**) has been computed with *Hyperchem* to determine its shape and dimensions. (**b**). The dotted spheres correspond roughly to the atomic sizes of C: *black*, H: *cyan*, N: *blue*, O: *red*, Zn: *green* given by the Van der Waals radii

carbon atoms of the phenyl rings (not the *tert*-butyl groups) approx. 1.6 nm. The phenyl groups are tilted by. ca. 70° relative to the porphyrin plane and the distance between the central carbon atoms of their *tert*-butyl groups is ca. 0.5 nm.

Upon self-assembly (aggregation) the oxygen groups and the *tert*-butylphenyl groups may slightly rotate out of these thermodynamic most stable positions in monomers, but this will not affect the assigned distances.

The molar mass of P2 is: 836.4 g/mol.

3.2 Synthesis

P2 has been synthesized by Balaban et al. starting from the fully synthetic compound: 10,20-bis-(3,5-di-*tert*-butylphenyl) copper porphyrin [1, 2] in four steps: (1) formation of P2-diacetyl copper porphyrin with a yield of 17 % by a reaction with acetic anhydride (Ac_2O) catalyzed by tin tetrachloride ($SnCl_4$) in carbon disulfide (CS_2), as reaction medium, a typical Friedel–Crafts solvent, (2) demetalation with trifluoracetic acid (TFA) + sulfuric acid (H_2SO_4) with a yield of 85 %, (3) reduction of one acetyl group to a hydroxyethyl group (monoreduction) to form the free base of P2 with a yield of 51 % by sodium borohydride ($NaBH_4$) in methanol and DCM, and (4) metallation with zinc by $Zn(OAc)_2$ in methanol and chloroform with a yield of 96 %.

The final purification step of P2 was the precipitation from a concentrated (a few tens mM) solution/dispersion in anhydrous DCM by mixing it with a major amount of anhydrous *n*-heptane, followed by centrifugation. After decanting the *n*-heptane with the residue of DCM the precipitated aggregates have been dried at a vacuum of about 0.1 mbar for several hours. The P2 stock was stored within a freezer at −20 °C, although these aggregates have been found to be very stable also at room temperature for several years.

3.3 Optical Properties

In this section, the optical properties of P2 monomers and aggregates have been investigated by absorption spectroscopy and fluorescence spectroscopy.

3.3.1 Absorption Spectroscopy

P2 was found to be the best soluble within tetrahydrofuran (THF), a solvent which combines polar and nonpolar characteristics. P2 is insoluble in pure polar and nonpolar solvents, like water or *n*-heptane respectively, dissolves however, well in nonpolar solvents, which contain a small amount of a polar component like water, methanol or THF. Other than DCM these polar compounds coordinate the porphyrin's central atom zinc [3] and thus hinder the self-assembly. The dissolution of the aggregates already starts upon the exposure of the solution to air, as this contains humidity, which dissolves also partly within *n*-heptane. Thus, to facilitate the self-assembly and avoid monomers the used solvents and glass ware were dried prior to use. Glass vessels and pipettes were dried in the drying oven at about 110 °C and the solvents DCM and *n*-heptane were refluxed over calcium hydride (CaH_2) and molten sodium, respectively, to remove the water impurities of the commercially available solvents. Forcing aggregation from monomeric solutions, that is, with hindered self-assembly, by drying-out the solvent, leads to the formation of amorphous aggregates, which show a broadened monomeric absorption spectrum. J-aggregates instead grow within anhydrous dichloromethane (DCM, CH_2Cl_2) either onto exposed solid surfaces, like the walls of the vessel, or precipitate and settle to the bottom within hours or days, most probably depending on the remaining water content, which is difficult to keep out completely. By the addition of anhydrous *n*-heptane the precipitation of J-aggregates can be accelerated up to seconds, depending on the ratio of *n*-heptane: DCM. In small vessels, the J-aggregates tend to grow onto the vessel walls, whereas in bigger ones they rather precipitate and settle to the bottom, because of the larger volume to wall surface ratio, but this depends also on the cuvette material (glass or quartz) and its content of adsorbed water.

In order to determine the exact band positions some absorption spectra of P2 have been also fitted by the following Gaussian function:

$$y = y_0 + \frac{A}{w\sqrt{\Pi/2}} e^{-2\frac{(x-x_c)^2}{w^2}}$$

where y_0 is the baseline, A is the area, w the full width at half maximum (FWHD) and x_C the center position of each band (peak).

Typical absorption spectra are plotted over wavelength, but for the fitting they needed to be plotted over a linear energy scale. In order to keep the absorbed spectral energy density constant, corresponding to the area under the graph, the

Absorbance needed to be corrected for a linear energy scale by multiplication with the following conversion factor:

$$A(E) = A(\lambda) \cdot \frac{\lambda^2}{\mathrm{h} \cdot \mathrm{c}} \text{ derived from: } \int A(E) \cdot dE = \int A(\lambda) \cdot d\lambda \rightarrow A(E)$$
$$= A(\lambda) \cdot \left(\frac{dE}{d\lambda}\right)^{-1}$$

where *A(E)* is the absorbance over a linear energy scale, *A(λ)* the absorbance measured over a linear wavelength scale, λ the wavelength, h the Planck's constant and c the speed of light.

All fittings were performed automatically by the software *Origin 6.1*, after defining the expected number of peaks. The center positions (xc) of the fitted absorption bands, listed next to each fit, can be easily converted into wavelengths by the following reciprocal conversion factor:

$$1239.84\text{nm} \approx 1240\text{nm} \mathrel{\widehat{=}} (1\text{eV})^{-1}$$

In pure THF P2 dissolves completely to monomers, as the self-assembly is hindered due to the coordination of the central atom zinc by the ethereal oxygen atom of THF. P2 solutions in THF appear pink to the eye and are clear, as they do not contain light scattering aggregates. Thus, for these absorption measurements no integrating sphere was needed.

The fit matches well the spectrum within the range of the main Q and B bands. The Soret band of P2 monomers in THF lies at 432 nm (2.871 eV), the Q bands $Q_x(1,0)$ and $Q_y(0,0)$ at 560 nm (2.214 eV) and 600 nm (2.066 eV), respectively. Additional information including the other bands, their areas and FWHM values are summarized in Fig. 3.2. The blue-shifted shoulder of the Soret band at 415 nm (2.988 eV) is supposed to be the vibronic transition B(1,0). Vibronic interactions are also responsible for the broad absorption within the ultraviolet (UV) wavelength range of porphyrins. Due to this broad absorption an uncertainty about the number of bands and their positions remains within this region. The band around 4.0 eV (310 nm) is supposed to have its origin in the aromatic phenyl rings, which are not conjugated together with the aromatic system of the porphyrin, due to their tilt of about 70° relative to the porphyrin plane.

In contrast to pure solutions of P2 monomers within THF, P2 aggregates appear not pink but green to the eye as dry deposits and in dispersions. In commercial *n*-heptane, which contained a small amount of water, assigned to $\leq$0.01 %, P2 could be dispersed by the application of ultrasound. After drying-out the solvent the following absorption spectrum of the cuvette walls was recorded of the macroscopically homogenous green film on the cuvette walls, as shown in Fig. 3.3. The fit gives again an estimate of the corresponding absorption bands.

Due to the red-shifted Soret band, visible as shoulder around 485 nm (2.554 eV) and the red-shifted Q bands at 585 nm (2.119 eV) and 633 nm (1.958 eV), the absorption spectrum comprises a much broader range of wavelengths compared to

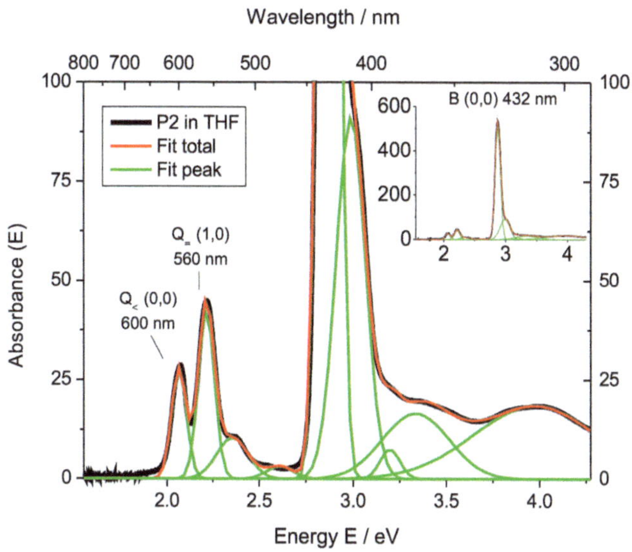

Model: Gauss
Chi^2/DoF = 3.52509
R^2 = 0.99948

y0	0	344
xc1	2.06601	30.00071
w1	0.08027	30.00149
A1	2.807	30.04444
xc2	2.21409	30.00088
w2	0.08626	30.00211
A2	4.63264	30.24715
xc3	2.35528	30.0094
w3	0.16891	30.02505
A3	2.18758	30.32135
xc4	2.61147	30.01515
w4	0.1441	30.03058
A4	0.57166	30.11917
xc5	2.87055	30.00012
w5	0.08426	30.00029
A5	54.55453	30.57163
xc6	2.98786	30.0028
w6	0.15607	30.00586
A6	17.83773	31.01592
xc7	3.19821	30.00785
w7	0.10562	30.02485
A7	0.98615	30.59914
xc8	3.33583	30.02896
6	0.354	30.06465
A8	7.35909	32.07307
xc9	3.97513	30.01733
61	0.65599	30.06489
A9	15.19414	31.39824

Fig. 3.2 Absorption spectrum of P2 monomers in THF with fitting. **a** Typical absorption spectrum of P2 monomers in a solution of anhydrous tetrahydrofuran (THF) (dark line) with a fit (*red*) consisting of nine Gaussian peaks (*green*). The inset shows the full spectrum. All fitting parameters were free for the automatic fitting algorithm with *Origin 6.1*. The center positions of the three main bands (xc1, 3, and 5) were converted also into nm and assigned in the graph

the monomeric one. A part of the Soret band remained obviously not red-shifted, but broadened around 433 nm (2.86 eV). It may be subdivided also into B(1,0) and B(0,0), but such details can not be resolved within this broad spectrum. The shoulder around 425 nm (2.92 eV) is supposedly not a blue-shifted part of the Soret band, but rather a red-shifted band, which appeared within the monomeric spectrum around 3.34 eV. Disregarding the shift induced by the solvent effect of THF in the monomeric spectrum, which is in the range of a few tens of meV, the red-shift of the Soret band was about 0.32 eV and that of both Q bands about 0.1 eV, compared to the monomeric bands in THF. The band around 4.75 eV might be blue-shifted, but this region was difficult to fit. This band is assumed to belong to the phenyl rings, which are tilted by about 70° relative to the porphyrin plane, hence couple, if at all, differently compared to the head-to-tail coupling of the porphyrin's transition dipole moments. In P2 J-aggregates the Q band with the lower energy ($Q_{Y,J}$) has a higher intensity than the higher energetic Q band ($Q_{X,J}$), in contrast to the monomeric spectrum where Q_X has a higher intensity than Q_Y. Despite the obviously good Gaussian fit, it should be kept in mind that due to the broad absorption around the Soret band it is difficult to determine the exact number of superposing bands and their exact positions.

Figure 3.4 compares the absorption spectra of P2 within different solvents on a conventional linear wavelength scale.

The absorption spectrum of P2 within anhydrous DCM resembles a combination of both spectra, that of monomers and of J-aggregates. The dispersed

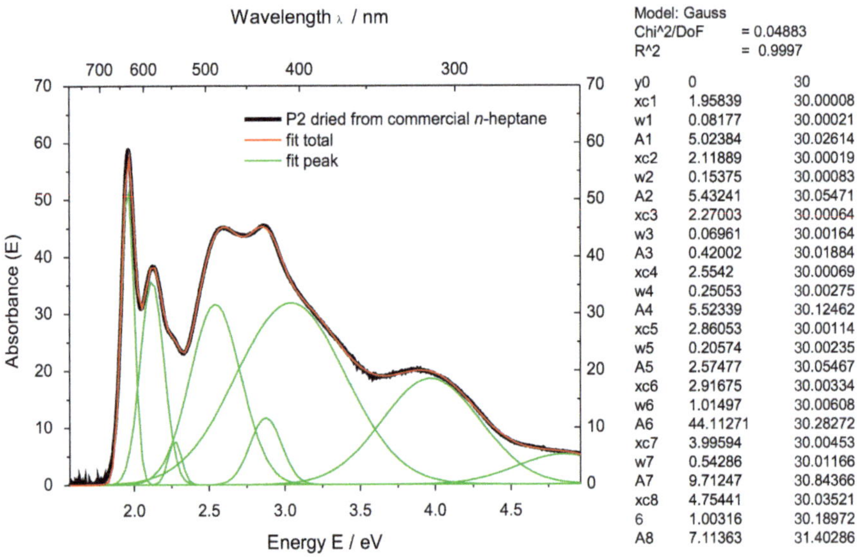

Fig. 3.3 Absorption spectrum of dried P2 J-aggregates with fitting. P2 J-aggregates were deposited from a dispersion in commercial *n*-heptane onto the walls of a quartz cuvette, by letting the solvent dry-out completely. The fitting was done with *Origin 6.1*, while all parameters were free to vary, The fitting parameters (xc: band center position in eV, w: FWHD in eV, A: area in eV ·Abs) for the eight bands, which have been found to fit the best for this spectrum, are listed on the right-hand side

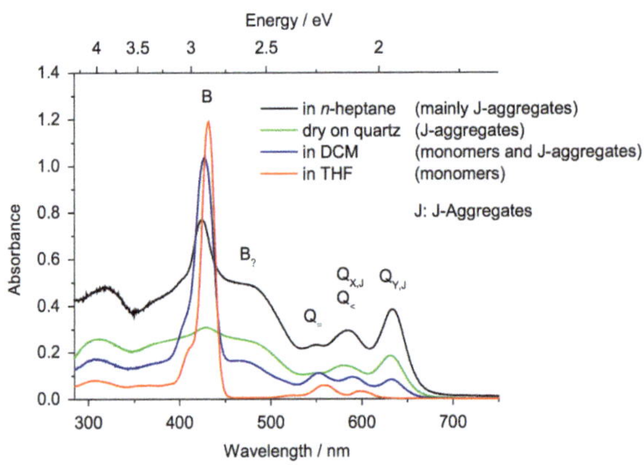

Fig. 3.4 Absorption spectra of P2 within different solvents and as dry deposits on quartz. The previous absorption spectra shown in Fig. 3.3 and 3.4 of P2 within anhydrous THF and as dry deposits on quartz are here plotted against the linear wavelength scale (*red* and *green*, respectively) in comparison with the absorption spectra of P2 dissolved/dispersed within commercial *n*-heptane (dark) and anhydrous DCM (*blue*). In anhydrous DCM. The *red*-shifted bands of J-aggregates are labeled with the index J. The absorption spectra of all samples, except that of the solution in THF, were measured within an integrating sphere

aggregates are supposed to be in a dynamic equilibrium with the monomers and the ratio between the band intensities of monomers and J-aggregates is very sensitive to the temperature, the concentrations of P2 and solvent impurities, like water or *n*-heptane [4].

Beside the band-shift due to the exciton coupling, the solvent does also affect slightly the band positions within a few nanometers due to the solvation effect of polar compounds which coordinate the central zinc atom. Thus, THF induces a small red-shift to all porphyrin bands, like the Soret and the Q-bands. The band at approx. 4.0 eV instead, is obviously not affected by this red-shift, compared to the other solvents. This agrees with the assumption that this band does not correspond to the conjugated system of the porphyrin, but to that of the phenyl rings, which are non-polar, thus not coordinated by polar compounds (Fig. 3.5, Table 3.1).

The corresponding absorption band maxima are summarized in the following table:

The small hump at the monomeric Soret band position of the P2 dispersion within the commercial *n*-heptane indicates a small amount of monomers, probably caused by the water impurity. That this band at the monomeric Soret band position belongs to separated monomers, and not to a change in the aggregate structure, could be shown in the following experiment in Fig. 3.6. There the J-aggregate growth has been induced onto the cuvette walls by injecting a concentrated anhydrous P2-DCM solution into a major amount of anhydrous *n*-heptane. Due to the adsorption of the J-aggregates onto the cuvette walls, the aggregates could be

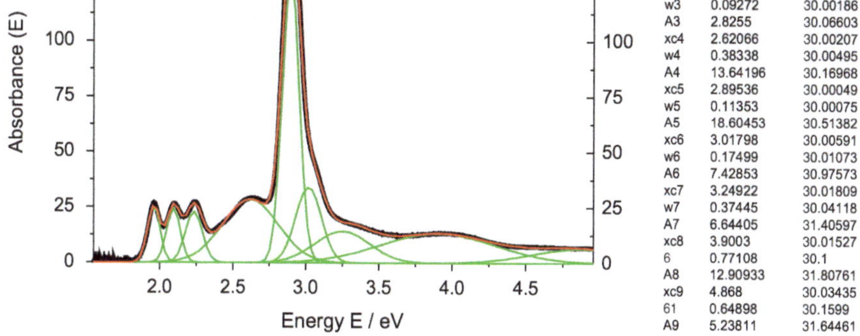

Fig. 3.5 Absorption spectrum of P2 in anhydrous DCM with fitting. This graph shows the fitting of the same absorption spectrum of P2 in anhydrous DCM, which was shown in Fig. 3.5 on the linear wavelength scale, again after converting the spectrum to a linear energy scale

Table 3.1 Absorption bands of P2 in different solvents. Absorption band maxima in nm, estimated from the original spectra on a linear wavelength scale, together with the fitted values in brackets, which were back converted to the nm unit, J: J-aggregates

Solvent	B	B_J	Q_X	$Q_{X,J}/Q_Y$	$Q_{Y,J}$
n-heptane	425 (433)	(485)	550 (547)	584 (585)	633 (634)
No solvent	429 (433)	(485)	(546)	580 (585)	631 (633)
DCM	427 (428)	(473)	553 (555)	589 (592)	632 (633)
THF	432 (432)	–	559 (560)	598 (600)	–

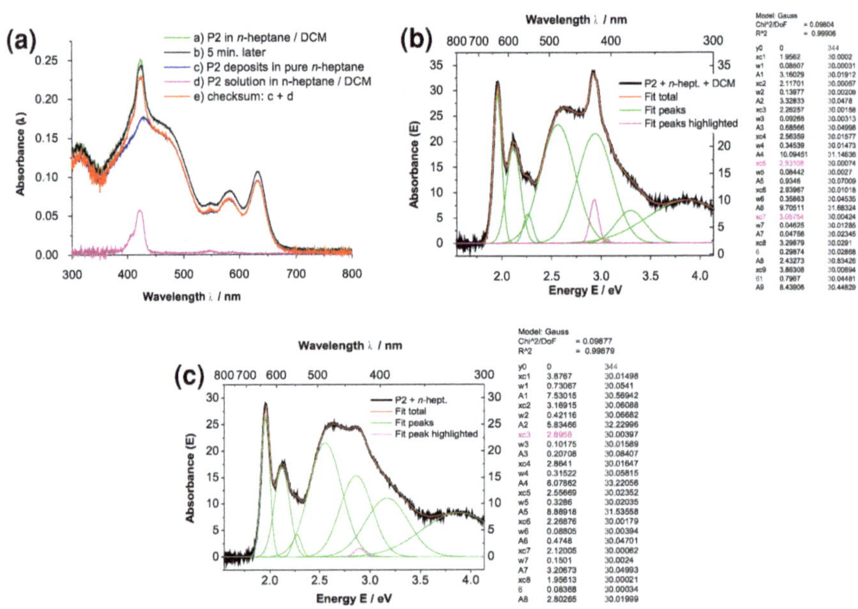

Fig. 3.6 Absorption spectra of coexisting P2 monomers and J-aggregates within DCM+*n*-heptane. **a** 20 μl of a 2.1 mM solution of P2 in anhydrous DCM were injected into a quartz cuvette (4 mm path) containing 1-ml anhydrous *n*-heptane (the *n*-heptane to DCM ratio was 50:1). Ca. 1 min after mixing the solvents the absorption spectrum was recorded within an integrating sphere (dark). A second measurement after 5 min revealed almost no change (*green*), thus the precipitation was completed within 1 minute, already before the first measurement. The liquid phase from this cuvette was exchanged by pure anhydrous *n*-heptane to measure the absorption of the deposits on the cuvette walls (*blue*). The absorption spectrum of the liquid phase was recorded within a clean cuvette (*magenta*), and revealed the presence of only P2 monomers. A calculated spectrum summarizing the last two spectra proved that no qualitative change occurred to the P2 J-aggregate spectrum by exchanging the solvent (*red*). Small differences in intensities can be attributed to a slightly different cuvette positioning within the integrating sphere. **b** Fit of the *green* trace in (a) (after conversion to a linear energy scale). All parameters were free to vary for the fitting with *Origin 6.1*. The bands highlighted in *magenta*, were found to have similar shapes as those measured within the liquid phase. **c** Fit of the J-aggregate deposits solvated with pure *n*-heptane (a, *blue*) proving, that almost no monomeric residue remained (*magenta*). All parameters were free to vary for this fitting, too

separated completely from the dissolved monomers which were present within the liquid phase.

By adding anhydrous *n*-heptane to an anhydrous P2-DCM solution (dispersion) within a quartz cuvette (*n*-heptane : DCM ratio = 50 : 1) the precipitation of P2 J-aggregates occurred predominantly onto the quartz cuvette walls, resulting in a macroscopically homogeneous film of J-aggregates. The liquid phase was then poured out into a second cuvette. The cuvette was then filled with pure anhydrous *n*-heptane and the absorption measurement (Fig. 3.6a) showed no significant difference to the absorption spectrum of the dry J-aggregates, measured before the solvation by *n*-heptane (not shown). In these spectra the hump at the monomeric Soret band position was much smaller compared to that one, where the DCM from the precipitation was still inside. This DCM - *n*-heptane mixture was measured after the precipitation separately within a clean cuvette and showed only the missing Soret band fraction, hence revealing the presence of only monomers. A calculated spectrum, obtained by summarization of these two spectra (monomers within the liquid phase and J-aggregates within the stained cuvette solvated by pure *n*-heptane), was qualitatively identical to the spectrum of both phases together within the cuvette where the precipitation was induced. This proved that the narrow monomeric band neither arises from monomers attached to the J-aggregates, nor from a change of the aggregates' structure. The finding of two similar bands in the fit (Fig. 3.6b, magenta) proves the quality of the fitting algorithm and shows that also a not shifted Soret band remains within the aggregates, which is broader than that of the monomeric Soret bands. This broadened not red-shifted Soret band within J-aggregates has also been found in the fit of the absorbance of the separated J-aggregate fraction (deposits on the cuvette walls) (Fig. 3.6c), where the monomeric Soret band residue (magenta) vanished almost completely. In spite of the good fit results, it has to be kept in mind, that the broad absorption around the Soret band is difficult to fit precisely and may be composed also of more bands than assumed in this model.

The following absorption spectra show the dissolution kinetics of these J-aggregate deposits of Fig. 3.6c after filling the cuvette again with anhydrous DCM. Within minutes to hours, the not shifted Soret and Q bands rose at 426 and 560 nm, respectively, while the red-shifted bands (Soret band shoulder around 470 nm and Q band at 532 nm) decreased (Fig. 3.7). This dissolution may also be an autocatalytic process like it has been encountered for the self-assembly of BChl *c* [5].

3.3.2 Fluorescence Spectroscopy

The distinction between P2 monomers and J-aggregates is also possible by fluorescence spectroscopy. In solutions/dispersions, containing monomers and aggregates, both species can be selectively excited within their specific absorption bands. As could be seen in the previous experiment in Fig. 3.6a and b, within the

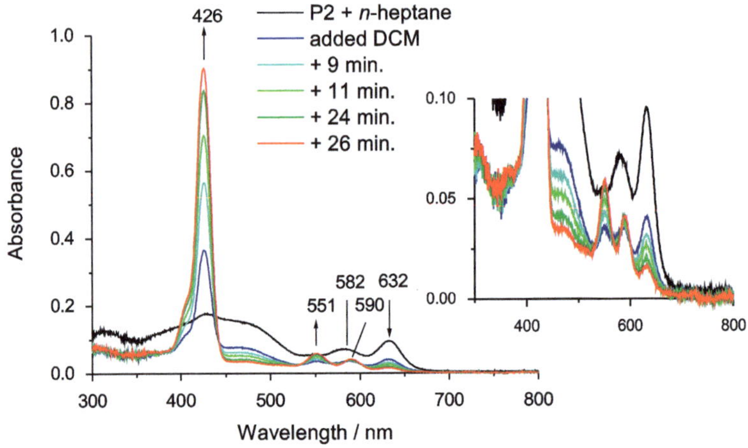

Fig. 3.7 Absorption spectra monitoring the dissolution of P2 J-aggregates in DCM. The same cuvette, with the P2 J-aggregates on the walls (dark) solvated in pure (anhydrous) *n*-heptane as shown in Fig. 3.6a, c, was filled with anhydrous DCM and shook gently. This initiated a dissociation process of the J-aggregates to monomers leading to a steady rise of the Soret band at 426 nm and of the monomeric Q band at 551 nm, whereas the *red*-shifted Soret and Q band around 470 nm and at 632 nm, respectively, decreased at the same time during 70 min (*blue*–*red*). The inset shows a magnification

solvent mixture of anhydrous *n*-heptane and a small amount of anhydrous DCM, P2 is not only present as J-aggregates, but also partly resides dissolved as monomers. Such a solution/dispersion has been used to selectively excite monomers and J-aggregates at 420 and 480 nm, respectively. After recording the absorption and fluorescence spectra a few drops of methanol were added to this solution in order to disassemble the aggregates to monomers. Then the fluorescence measurements have been repeated. The results are shown in Fig. 3.8.

The typical J-aggregate absorption spectrum changed almost to a pure monomeric one upon the addition of methanol (Fig. 3.8a). The slight red-shift of this monomeric absorption spectrum, which is even enhanced in the fluorescence spectrum, is caused by a complexation of the porphyrin's central zinc atom by methanol. Upon the excitation at the J-aggregate's absorption band at 480 nm the fluorescence intensity was significantly lower, than that measured upon excitation of the monomers at 420 nm. Within the predominantly monomeric solution, containing methanol, there is no significant qualitative difference in the fluorescence spectra at both excitation wavelength (Fig. 3.8b, blue, green). Without the methanol instead, both fluorescence spectra differ also qualitatively from each other (dark, red). There the main fluorescence band of the monomers at 592 nm almost disappeared and the band around 713 nm, which was a small shoulder upon excitation of the monomers, developed to a distinct band upon excitation within the aggregate's absorption band. Thus, we assume, that these two fluorescence bands, at approx. 647 and 713 nm, origin from the J-aggregates and the bands at 592 and 647 nm origin from the monomers' decay.

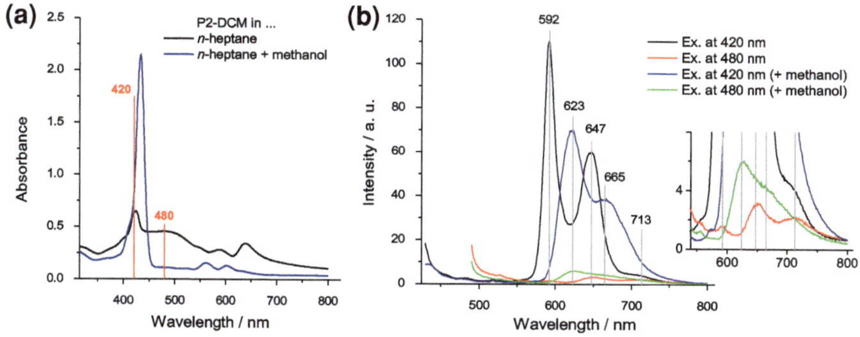

Fig. 3.8 Fluorescence spectra of P2 within *n*-heptane with and without methanol. A dispersion of P2 within anhydrous DCM has been injected into a cuvette with anhydrous *n*-heptane to induce precipitation of P2 J-aggregates. After recording the absorption (**a**, dark) and fluorescence spectra (**b**, dark and *red*) 2–3 drops of methanol were added into the cuvette followed by a gently shake and repeat of the measurements (*blue*, *green*). The inset shows a magnification of the bottom part of the fluorescence spectrum. The tail of the absorption spectrum of the aggregate dispersion (**a**, dark) is an artifact caused by light scattering at the J-aggregates, because no integrating sphere was used here. Reprinted with permission from Chappaz-Gillot C, Marek PL, Blaive BJ, Canard G, Bürck J, Garab G, Hahn H, Jávorfi T, Kelemen L, Krupke R, Mössinger D, Ormos P, Reddy CM, Roussel C, Steinbach G, Szabó M, Ulrich AS, Vanthuyne N, Vijayaraghavan A, Zupcanova A, Balaban TS (2012) J Am Chem Soc 134:944-954. Copyright 2012 American Chemical Society

The shoulder of the monomer's fluorescence at 713 nm may be caused by an energy transfer from the monomers to the J-aggregates. Another possibility would be a decay from the monomeric LUMO into the second excited vibrational level of the monomeric HOMO, whereas the band at 647 nm is supposed to be the decay into the first excited vibrational level and the band at 592 nm is the decay into the ground state of the monomeric HOMO. The almost disappearing of the monomeric band at 592 nm upon excitation of the aggregates at 480 nm and the development of the fluorescence band at approx. 713 nm can be explained by a red-shift of both (or the possible three) monomeric bands by 0.18 eV each. The possible third monomeric band cannot be resolved in the red-shifted J-aggregate spectrum, due to the low intensity level. This obvious red-shift in the fluorescence spectrum asserts, that the J-aggregates have been indeed selectively excited and that they fluoresce in the solid-state, what is a good indication for a long-lived excited state.

Time resolved fluorescence measurements have been done by cooperation with Jędrzej Szmydkowski (group of Heinz Kalt, KIT) in order to investigate the lifetime of the excited states, which may allow to estimate the exciton diffusion time within the aggregates. A similar solution/dispersion of P2 as in the previous experiment has been prepared within *n*-heptane to selectively excite the monomers and J-aggregates by different wavelengths. The precipitation of J-aggregates has been induced by injecting a small amount of an anhydrous P2-DCM solution into a major amount of *n*-heptane with a very low water content ($\leq$0.005 %). The monomers were excited by a pulsed laser at 430 nm and aside the monomeric absorption band the J-aggregates

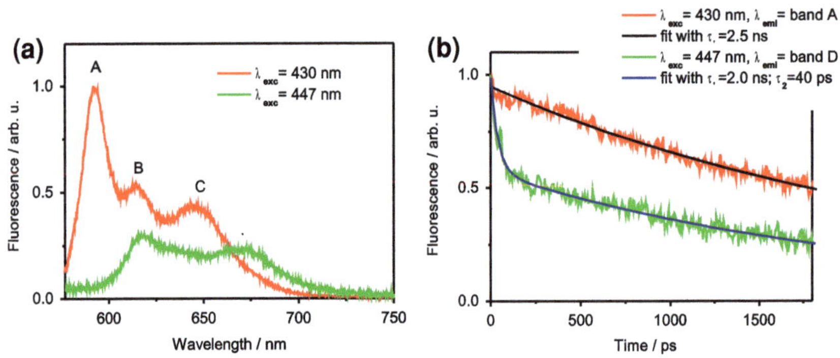

Fig. 3.9 Wavelength- and time-resolved fluorescence of P2 monomers and J-aggregates. P2 was dissolved within anhydrous DCM and injected into commercial dry *n*-heptane (≤0.005 % water) to induce the precipitation of J-aggregates. After the pulsed excitation with a laser at 430 nm (*red*) and at 447 nm (*green*) the decay of the fluorescence signal was measured wavelength- (**a**) and time-resolved (**b**) by a CCD detector. The time-resolved fluorescence measurements of the fluorescence band denoted with "A," which originated from an excitation at 430 nm (**a**, *red*), showed a monoexponential decay, where the mean lifetime could be fitted to 2.5 ns (**b**, dark). The decay of the fluorescence band "D" (*green*) could be fitted by a biexponential function, corresponding to two mean lifetimes of 2.0 ns and 40 ps (*blue*). Reprinted with permission from Huijser, A.; Marek, P.L.; Savenije, T.J.; Siebbeles, L.D.A.; Scherer, T.; Hauschild, R.; Szmytkowski, J.; Kalt, H.; Hahn, H.; Balaban, T.S., J. Phys. Chem. C, 2007, 111, 11726-11733. Copyright 2007 American Chemical Society

were excited at 447 nm. The time- and wavelength-resolved measurements are shown in Fig. 3.9.

The excitation within the Soret band of monomers at 430 nm gave a fluorescence spectrum with three bands, labeled as A, B and C (Fig. 3.9a: red). Band A and C are expected to be the two main fluorescence bands of P2 monomers, whereas the band B most probably originates from the complexation of P2 monomers with water, because a similar red-shift was caused by the monomer-methanol complex known from Fig. 3.8b. The excitation wavelength of 447 nm was intended to excite only the J-aggregates, but the biexponential fluorescence decay in the time-resolved measurement suggests the excitation of two different species (Fig. 3.9b). Whereas the excitation of monomers at 430 nm led to a monoexponential decay of the fluorescence band A (Fig. 3.9b: red, black), the decay of the fluorescence at the wavelength D, after excitation at 447 nm, implies the existence of two excited state lifetimes, affirmed by the good fitting with a biexponential function (blue). As the fluorescence band A was only present after excitation within the monomer's absorption band, the fitted mean lifetime of 2.5 ns is expected to originate from the monomers. The fluorescence band D of the excited J-aggregates instead, has an overlap with the fluorescence band C, which occurred by exciting the monomers. Thus, the two decay times at the wavelength D of 2.0 nm and 40 ps are supposed to correspond to the fluorescence of monomers and J-aggregates, respectively. The question that arises here is how could the

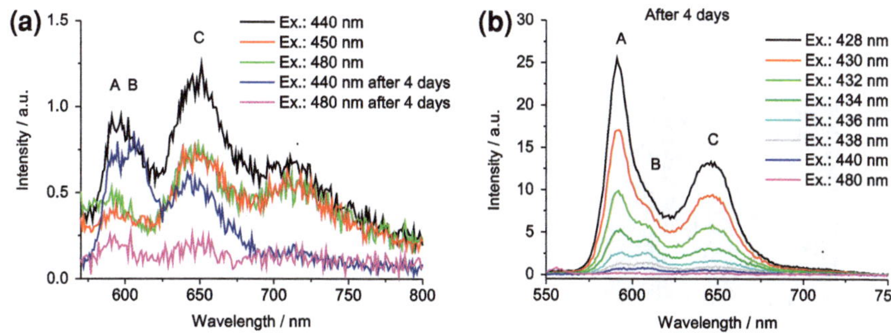

Fig. 3.10 Fluorescence spectra of P2 within commercial *n*-heptane as prepared and after 4 days. An anhydrous P2-DCM solution was injected into a cuvette with a major amount of commercial *n*-heptane to induce the precipitation of J-aggregates. The fluorescence of this solution/dispersion was measured immediately after preparation (**a**: dark, *red*, *green*) and after 4 days storage within the sealed cuvette (*blue*, *magenta*). The two last spectra (*blue*, *magenta*) are also shown in (**b**) together with a series of fluorescence spectra measured at different excitation wavelengths of this 4 days old solution. The curves in (**a**) are very noisy, due to the low intensity and a very fast scan rate of 10 nm/s of the emission wavelengths. In addition the resolution was limited by the slit bandwidth to 5 nm for the excitation and the emission wavelength

monomers got excited at 447 nm, too, when the experiment in Fig. 3.6a and b showed, that the absorbance of the residual monomers in solvent mixtures of *n*-heptane and DCM is almost zero at this wavelength. The answer may give the unexpected band B, which is at a similar wavelength as the monomer-methanol complex of the previous experiment and may be attributed to a similar complex with water, which was contained to a small amount ($\leq$0.005 %) within the commercial *n*-heptane. The wavelength for exciting the aggregates could not have been set to higher values as 447 nm, because this was the limit of the used optical parametric oscillator, which tuned the laser wavelength. At this wavelength, a red-shifted monomer complex with water could be excited beside the J-aggregates, too. Thus, the longer excited state lifetime of 2.0 ns supposedly corresponds to the decay of the monomer-water complex, whereas the shorter lifetime of 40 ps is attributed to the lifetime of the J-aggregates.

The fluorescence of a similar solution/dispersion of P2, aggregated from an anhydrous DCM solution/dispersion in commercial *n*-heptane ($\leq$0.005 %), has been been measured wavelength-resolved freshly after preparation and after 4 days storage at room temperature (Fig. 3.10).

At the excitation wavelength of 440 nm all four bands (A-D) are present within the fluorescence spectrum (Fig. 3.10a, dark). At the excitation wavelength of 450 nm the monomeric band A, as well as the band B of the supposed monomer-water complex, decreased significantly and didn't decrease further at the excitation wavelength of J-aggregates at 480 nm (red, green). After storage of the sealed cuvette for four days at room temperature the fluorescence band of the aggregate (D) vanished almost completely and only the monomeric bands (A, B, C) remained at the excitation wavelength of 440 nm (blue). The additional vanishing of the

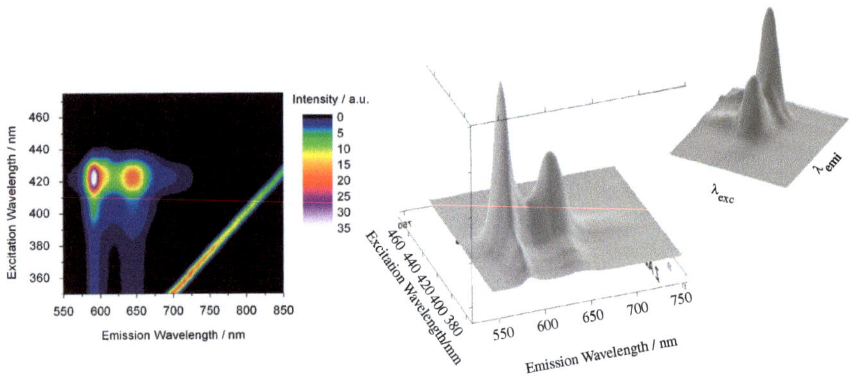

Fig. 3.11 3D fluorescence spectrum of P2 monomers within commercial *n*-heptane. P2 was dissolved/dispersed within anhydrous DCM and given into a major amount of commercial *n*-heptane (water content $\leq$0.005 %). After 4 days sealed storage, this 3D fluorescence spectrum was recorded. This spectrum, which shows mainly the fluorescence of P2 monomers, is shown in two different styles: the contour plot (left-hand side) and the topographic view from two sides with the fluorescence intensity on the vertical scale. The line within the contour plot from the bottom to the top right corner is an artifact of the spectrometer

monomeric fluorescence bands at the excitation wavelength of 480 nm proves that at this wavelength no monomers can be excited, just exclusively the J-aggregates.

Although the absorption spectrum of this solution/dispersion shows the presence of monomers and J-aggregates (not shown here) the latter lost their ability to fluoresce almost completely within this liquid. As the chemical stability of P2 within J-aggregates is very high, what will be shown later, this absence of fluorescence may be due to a change in aggregate morphology. A possible change at this conditions and within this period of time may be the growth of small dispersed J-aggregates to bigger ones or the disordered agglomeration of small aggregates to larger clusters which settled onto the cuvette walls. This can mean, that the trap density, at which excitons quench, increases with the growth of the aggregates or the agglomerates.

Figure 3.10b shows the three dimensional fluorescence spectrum of the four days old solution in its original sealed cuvette. The decrease of the band A and the evolution of the band B can be seen in the smaller topographic view (top right corner) at higher excitation wavelengths. This measurement is an excerpt from a three-dimensional (3D) fluorescence spectrum of this solution, which is shown in Fig. 3.11.

3.4 Processing and Morphology of the Aggregates

In this section, the morphology and structure of P2 aggregates, deposited from DCM dispersions, has been investigated by SEM, TEM, and XRD.

3.4.1 Drop-casting from Dichloromethane (DCM)

DCM is an often used solvent for self-assembling porphyrins, because it has a high potential to dissolve or disperse them without affecting their self-assembly. P2 J-aggregates can be obtained with well-defined facets and a low content of an amorphous phase directly from anhydrous DCM solutions/dispersions or by the precipitation from mixtures of DCM and *n*-heptane. Drop-casting is an efficient way to coat a substrate with such aggregates simply by letting dry one or more drops of this dispersion on a substrate. For the naked eye such coatings look not very homogenous, compared to those prepared by spin-coating for instance, on a microscopic scale however, the deposits look very similar.

Figure 3.12 shows typical P2 aggregates after drop-casting from a 2.1 mM anhydrous DCM dispersion onto a silicon wafer. The resulting rod-like aggregates have a wide size-distribution, where only a small fraction of aggregates is thin enough to not exceed significantly the exciton diffusion length, which is assumed to be in the order of a few tens of nm.

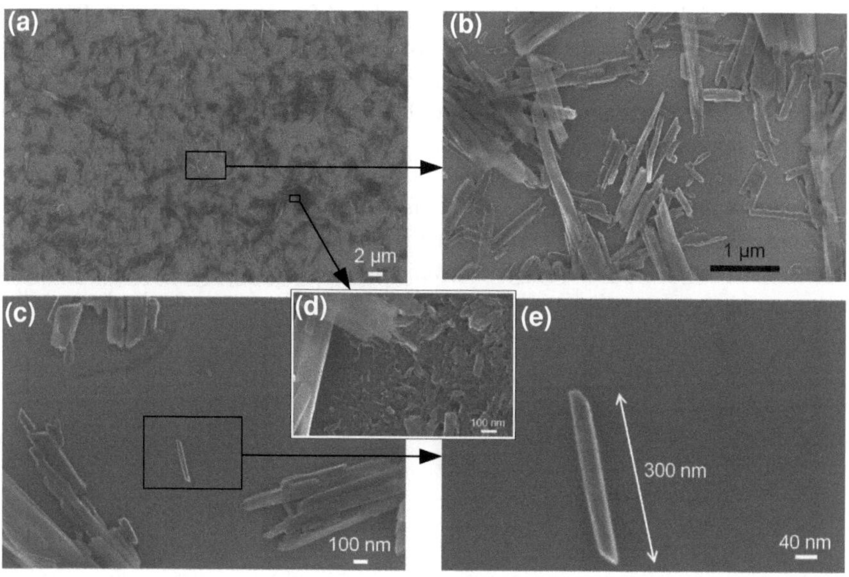

Fig. 3.12 SEM micrographs of P2 aggregates on silicon deposited by drop-casting from DCM. P2 has been dissolved within anhydrous DCM (CH_2Cl_2) to a concentration of 2.1 mM by sonicating it for 5 min and keeping the dispersion sealed for a few hours in the dark. A drop of this dispersion has been dried onto a smooth silicon wafer under ambient conditions. In-lens, EHT: 10 kV, WD: 4 mm, aperture: 30 µm. (**b**) [Marek PL, Hahn H, Balaban TS (2011) Energy Environ Sci 4:2366–2378] – Reproduced by permission of The Royal Society of Chemistry. (**e**) Reprinted with permission from Chappaz-Gillot C, Marek PL, Blaive BJ, Canard G, Bürck J, Garab G, Hahn H, Jávorfi T, Kelemen L, Krupke R, Mössinger D, Ormos P, Reddy CM, Roussel C, Steinbach G, Szabó M, Ulrich AS, Vanthuyne N, Vijayaraghavan A, Zupcanova A, Balaban TS (2012) J Am Chem Soc 134:944-954. Copyright 2012 American Chemical Society Ormos P, Reddy CMCopyright 2007 American Chemical

The overview in (a) with its magnification in (b) shows a typical P2 aggregate distribution, where some larger aggregates can be seen to be highly transparent for electrons. The tiny aggregate magnified in **e** reaches already a width of 40 nm, while its internal structure is apparently highly ordered, as can be assumed from the high aspect ratio of 7.5 and the straight facets, which are almost like in a conventional crystal.

The micrograph in (d) shows a magnification of a dark spot, of which only two were visible in the overview (a). These are agglomerates of very small aggregates, which did not have such regular shapes and such aspect ratios like the above described aggregate. They might have a more disordered internal structure, but are not amorphous as can by seen on their distinct facets.

3.4.2 Precipitation from DCM – n*-heptane Mixtures*

Similar P2 J-aggregates can be obtained also by precipitation from a concentrated P2 dispersion within anhydrous DCM when added into a larger amount of anhydrous *n*-heptane (n-C7), in which a substrate was placed for the coverage with P2 aggregates. By this technique the aggregates grow also on the substrate surface which is directed downward when the liquid phase has enough access to this surface, for example, when its not sticking tightly to the bottom of the vessel. The results of J-aggregates processed by this precipitation technique is shown in Fig. 3.13.

The P2 aggregates and their distribution differed not much from the morphology obtained by the drop-casting technique directly from anhydrous DCM (Fig. 3.12). The precipitation technique in *n*-heptane led to a macroscopically more homogeneous distribution of the aggregates over the whole substrate without significant macroscopic concentration gradients nor drying rings, in contrast to the drop-casting

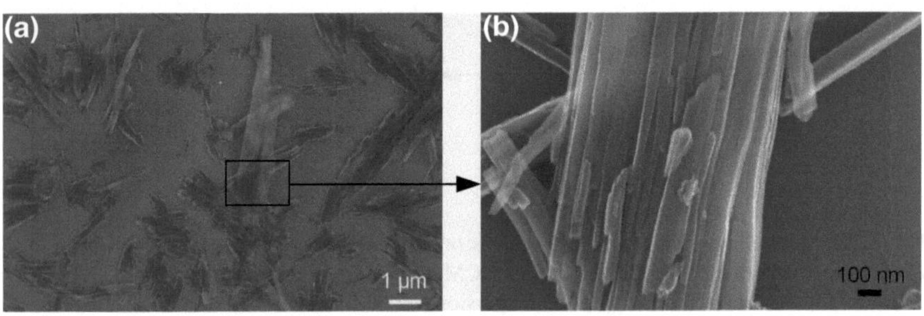

Fig. 3.13 SEM micrographs of P2 aggregates on silicon deposited by precipitation in n-C_7. 20 µl of the same 2.1 mM dispersion of P2 within anhydrous DCM as in Fig. 3.12 were injected into 1 ml of anhydrous *n*-heptane (n-C_7) containing an additional amount of 30 µl of anhydrous DCM. The aggregates precipitated completely within seconds and macroscopic homogeneous onto all walls of the vessel and the inside placed silicon wafer. In-lens, EHT: 10 kV, WD: 4 mm, aperture: 30 µm

technique. This technique is also less sensitive to impurities within the solvent, because the *n*-heptane is not allowed to dry completely onto the substrate, thus more impurities and monomers can stay within the remaining solvent. By the drop-casting technique also amorphous residues can be deposited on the substrate if the solvent was not pure enough. The microscopic morphology and size-distribution of the aggregates is similar for both deposition techniques. There might be a slight tendency that the rod-shaped aggregates, which were deposited by the precipitation technique, agglomerated slightly more, as can be seen in Fig. 3.13b.

3.4.3 Testing Other Solvents: Dichlorobenzene and Diethyl Ether

Takahashi et al. [3] 1,2-dichlorobenzene, also called *ortho*-DCB or *o*-DCB, does also not coordinate the central zinc atom of zinc porphyrins, similar to DCM, in contrast to, for example, alcohols. As *o*-DCB has a much higher boiling point of 180 °C, compared to that of DCM, being 40 °C, the self-assembly at elevated temperatures may differ from that at room temperature from DCM. We investigated the resulting aggregates deposited by evaporating some drops of P2 dispersed within commercial *o*-DCB at 110 °C (Fig. 3.14).

Fig. 3.14 SEM micrographs of P2 J-aggregates from a *o*-DCB dispersion. A 0.1-mM dispersion of P2 in 1,2-dichlorobenzene (*o*-DCB) was drop-cast onto an FTO substrate. The solvent was evaporated in a drying oven at ca. 110 °C. In-lens, EHT: 5 kV, WD: 5 mm, aperture: 20 μm

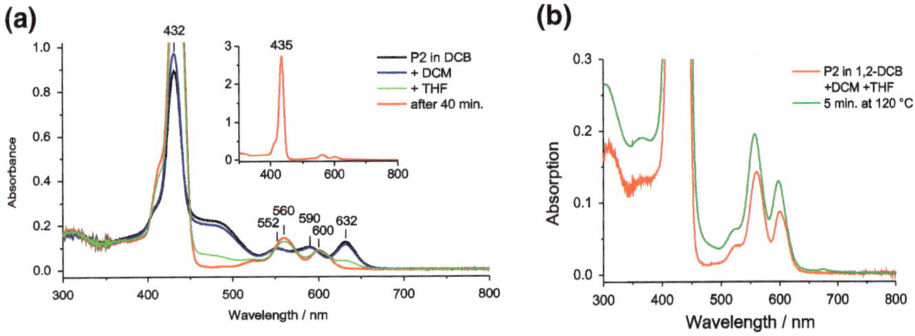

Fig. 3.15 Absorption spectra of P2 in *o*-DCB and with a trace of DCM and THF. **a** Into a 0.1 mM dispersion of P2 within commercial dichlorobenzene (*o*-DCB) 50-μl anhydrous dichloromethane (DCM) were added ($\approx$ 11 vol %) and afterward 50-μl anhydrous THF. Whereas the DCM did not change the absorption spectrum significantly, the addition of THF led to a disassembly of the P2 J-aggregates within about 40 min. **b** The solution was then heated up to 120 °C until the first bubbles arose and was kept at this temperature for about 5 min

The SEM micrographs revealed quite similar rod-shaped structures as known from DCM dispersions. The absorption spectrum of P2 dispersed within *o*-DCB, shown in Fig. 3.15, was also very similar to that in DCM, revealing a major fraction of J-aggregates. To this dispersion also a small amount of THF was added in order to see if it is possible to evaporate the THF, which has a boiling point of 66 °C, from the mixture with *o*-DCB, what might be indicated by a recovery of the self-assembly of P2.

After the addition of anhydrous DCM to the P2-*o*-DCB dispersion the absorption spectrum did not change significantly and still indicated mainly the presence of J-aggregates by the red-shifted Soret band (shoulder) and the red-shifted Q bands like that at 632 nm. After addition of THF, the J-aggregates disassembled almost completely within 40 min, what could be seen by the vanishing of the red-shifted Soret band and Q bands while the monomeric bands rose. The Soret band maximum slightly shifted from 432 to 435 nm due to the common solvent effect of the polar THF. The solution was has been annealed until it large bubbles started to rise at about 120 °C. After a few seconds, the boiling stopped and the temperature was maintained for 5 more minutes. The following absorption measurement of this solution showed almost no qualitative change compared to the former one containing THF in the *o*-DCB solution. The Soret band shifted partly back to 434 nm and the Q bands also by about 1–3 nm, showing that the THF was only partly removed again. The deposits from this annealed P2-THF-*o*-DCB solution were amorphous.

Also the deposition from the solvent diethyl ether was tested, which has a very similar molecular structure as THF, but is known to have very different properties, due to its acyclic conformation shielding the lone pairs of the oxygen [6]. The used commercial diethyl ether contained 0.2 % water and therefore has been distilled once to reduce the water content. The absorption spectrum of the P2 within diethyl ether was very similar to that within anhydrous DCM and *o*-DCB, revealing also

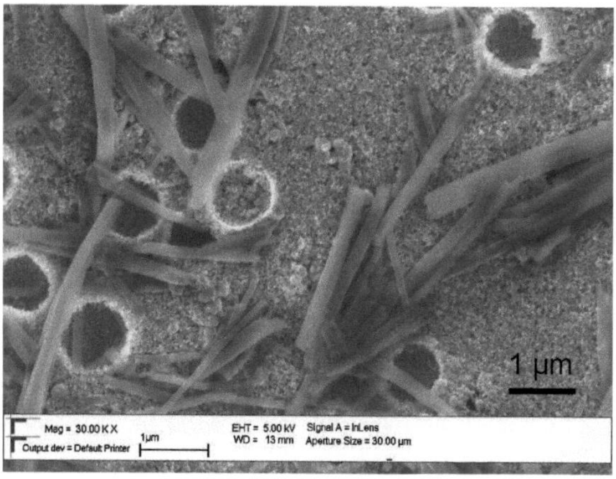

Fig. 3.16 SEM micrograph of P2 aggregates drop-cast from diethyl ether. P2 has been dispersed within commercial diethyl ether from *Merck* (0.2 % water), which was distilled once, within an ultrasonic bath and then drop-cast onto a mesoporous TiO_2 layer containing additional 1 μm-sized macropores. The solvent has been evaporated at 110 °C in air. In-lens, EHT: 5 kV, WD: 13 mm, aperture: 30 μm

predominantly J-aggregates. The aggregates have been drop-cast from this dispersion onto a mesoporous TiO_2 layer, which contained additional macropores. The J-aggregate morphology and size-distribution was also very similar to that obtained from anhydrous DCM dispersions, as can be seen in Fig. 3.16.

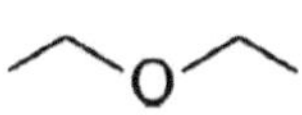

diethyl ether

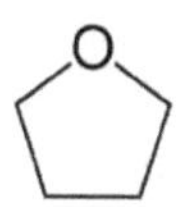

THF

3.5 Structure of P2 J-Aggregates

When the size or long-range order of crystals or well-structured aggregates is too short for single crystal analysis by X-ray diffraction (XRD), it might still be possible to determine their structure using electron diffraction within a TEM. It requires however, that the investigated aggregate resists the electron beam without structural damage. As the aggregates have a preferred orientation on the substrate it is necessary to record a series of electron diffraction patterns of a single aggregate at different angles, which it does not resist.

The aggregates have been directly deposited from a drop of an anhydrous P2-DCM dispersion onto a copper grid spanned by a thin amorphous carbon layer.

The TEM images revealed well-defined lattice fringes within the P2 aggregates as can be seen in Fig. 3.17.

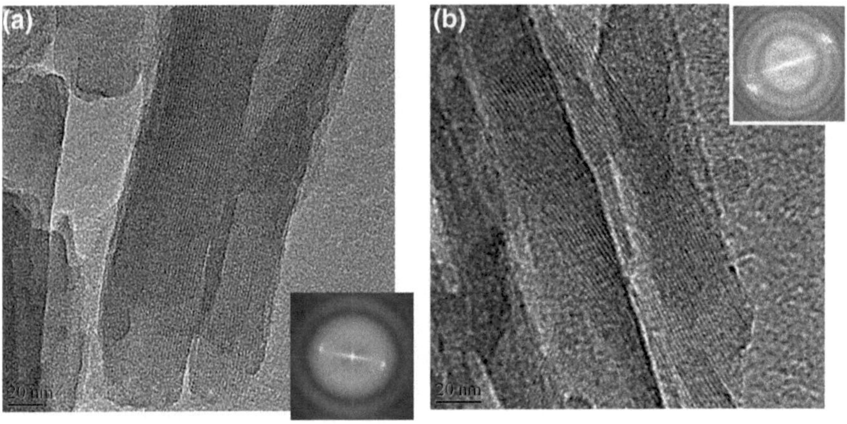

Fig. 3.17 TEM images of P2 aggregates with visible lattice fringes. P2 aggregates have been deposited from an anhydrous DCM dispersion onto an amorphous carbon layer. The accelerating voltage was 200 kV. The insets show the fast Fourier transforms (FFT) of the corresponding images. Reprinted with permission from Chappaz-Gillot C, Marek PL, Blaive BJ, Canard G, Bürck J, Garab G, Hahn H, Jávorfi T, Kelemen L, Krupke R, Mössinger D, Ormos P, Reddy CM, Roussel C, Steinbach G, Szabó M, Ulrich AS, Vanthuyne N, Vijayaraghavan A, Zupcanova A, Balaban TS (2012) J Am Chem Soc 134:944-954. Copyright 2012 American Chemical Society

In (a) the spacing in the projection of the lattice planes was 1.6 nm. In (b) the spacing in the projection of the lattice planes appeared from 1.6 nm in the bottom right corner to 1.8 nm in the top left corner. This gradient may indicate a twist in the aggregate. Unfortunately, these aggregates did not resist long enough the high energy of the electron beam, which had an accelerating voltage of 200 kV, to record a sufficient number of diffraction patterns at different angles, to derive a structural model. A lower accelerating voltage does not necessarily increase the stability during the TEM investigations and has the disadvantage of a lower resolution. The lattice fringes all along the aggregate's long axis prove a highly regular ordering of the aggregates, anyway. This pattern is very similar to that, which has been reported for BChl-chlorosomes [7] and also for other BChl mimics like the porphyrin no. 6 from Fig. 3.4 [8–10].

The fast Fourier transforms (FFT) of the lattice fringes reveal only reflexes in the inner circle showing, that the fine-structure of the aggregates was already damaged by the electron beam and only the largest lattice distance could be recorded. It has to be kept in mind, that the measured distances of the lattice fringes are only projections of the lattice planes, which can have a larger distance than 1.8 nm, what would be visible only in the right angle.

The structure of P2 aggregates has also been investigated by XRD. Therefore, P2 was deposited by drop-casting from a dispersion in anhydrous DCM onto different substrates, which have been investigated by XRD. Also the dry P2-powder has been measured without the treatment in any solvents. The diffraction peaks have been converted by the Bragg equation into distances of the periodic lattice planes:

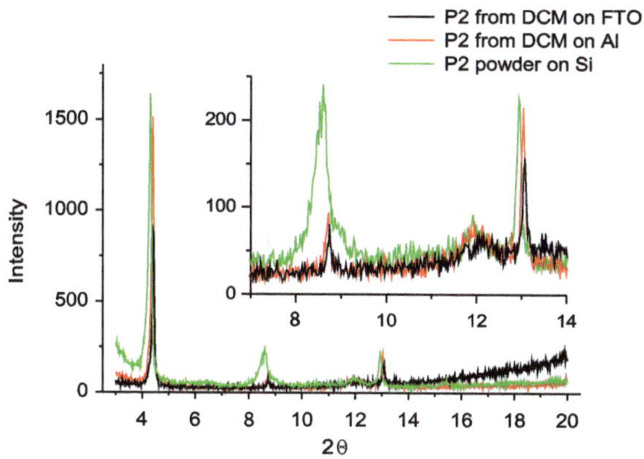

Fig. 3.18 XRD of P2 aggregates drop-cast from DCM and as dry powder. X-ray diffraction (XRD) patterns have been measured of P2 aggregates, which have been deposited by drop-casting of an anhydrous DCM dispersion onto fluorine-doped tin oxide (FTO, dark trace) and aluminum (Al, *red* trace). The diffraction pattern of the dry P2 powder from our stock has been measured on a silicon wafer (*green* trace). Diffractometer: D8, Bragg–Brentano geometry, 40 kV acceleration voltage, 40 mA current, 0.02° increment, scan speed 6, Göbel mirror and parallel plate collimator, Filter: Cu K_{β}

Bragg equation: $n\lambda = 2d\sin(\theta)$ with: $\lambda_{\text{Cu-K}\alpha\ (1\ \&\ 2)}$: 0.154 nm (Fig. 3.18).

The first peak of the P2 powder on Si appeared at 4.30° of 2Θ, what corresponds to a distance of the lattice planes of 2.05 nm. This is in accordance to the value reported by Balaban, which was 2.1 ± 0.05 nm [9]. Whereas the second peak at 8.60° (1.03 nm) may be a second order diffraction of the first one, the third peak at 12.95° would correspond to an additional lattice periodicity of 0.68 nm, what is twice the usual π–π interaction distance of 0.34 nm. The diffraction patterns obtained for P2, which has been deposited by drop-casting from an anhydrous DCM dispersion onto an FTO and Al substrate, were less broad and slightly shifted to higher 2Θ-angles. They correspond to lattice distances of 2.01, 1.01, and 0.68 nm. The powder from the stock was obtained by precipitating the aggregates from an anhydrous DCM dispersion in anhydrous *n*-heptane during centrifugation of the aggregates. Afterward they have been dried in vacuum within the centrifuge glass and collected by scratching them off from the glass. This mechanical stress may have perturbed the structure of the aggregates causing the broadened peaks of the dry P2 powder. These defects may have healed out by the additional dissolution/dispersion within DCM for the deposition by drop-casting. A similar lattice periodicity of 2.09 nm Pšenčík et al. has also been observed in the natural BChl-*c*-aggregates by measuring the XRD of chlorosomes [7].

By the XRD-measurements of P2 spin-coated from THF solutions no diffraction peaks could be detected. Not even when P2 was infiltrated from a THF solution into mesoporous TiO_2 layers, containing additional 1 μm sized macropores, followed by annealing these samples within anhydrous *n*-heptane. This confirms the absorption measurements, which showed that the mean aggregate structure of P2 deposits, infiltrated with THF and post-annealed in *n*-heptane, remained amorphous also when additional macropores were present within such layers. Hence, the apparently well ordered rod-shaped aggregates, which have been seen by SEM within some macropores belonged to a minor fraction of the aggregates. The average aggregate structure instead, which has been measured by absorption spectroscopy and XRD was amorphous. These aggregates probably stick within the mesopores and were not capable of being differentiated by SEM-investigations.

3.6 Thermal Stability

To determine if P2 is processable via evaporation or sublimation or if it decomposes in advance, a thermogravimetric analysis (TGA) with in situ quadrupole mass spectrometry (MS) of the gas phase was performed. The TGA is an analytical method to measure the change of the mass of a sample against the time and the applied temperature. The system used was equipped with a quadrupole mass spectrometer in order to analyze in situ the composition of the gas released by the sample (Fig. 3.19).

In a first TGA-MS measurement of P2 (2.593 mg) several ions have been automatically detected, most of them only at elevated temperatures above 350 °C. In a

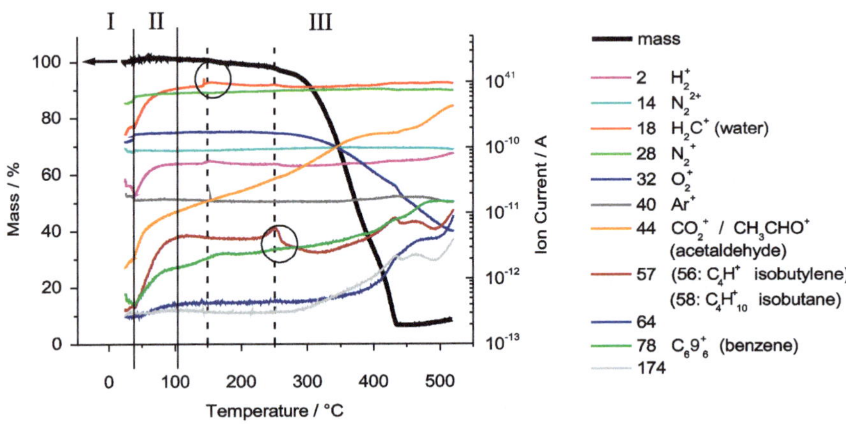

Fig. 3.19 Thermogravimetry with MS of P2 aggregates. The left axis shows the mass reduction of the P2 (J-aggregates) sample (100 % = 2.128 mg) which occurred during the thermogravimetry measurement and the right axis shows the ion current of the simultaneously performed MS in the evacuated chamber with $5 \cdot 10^{-6}$ mbar. A hold point for 1 h was set at 35 °C to desorb water (phase I) and afterward the heating rate was increased to reach its maximum of 4.8 K/min after 40 min at ca. 105 °C(II), which was then kept constant up to 520 °C(III). Reprinted with permission from Chappaz-Gillot C, Marek PL, Blaive BJ, Canard G, Bürck J,Garab G, Hahn H, Jávorfi T, Kelemen L, Krupke R, Mössinger D, Ormos P, Reddy CM, Roussel C, Steinbach G, Szabó M, Ulrich AS, Vanthuyne N, Vijayaraghavan A, Zupcanova A, Balaban TS (2012) J Am Chem Soc 134:944-954. Copyright 2012 American Chemical Society.

second run another batch of P2 (2.128 mg) was measured, tracking the signals of the previously detected ions, from the beginning on. In the second run also a hold point was set at 35 °C in vacuum for 1 h in order to desorb water, which was detected in a higher quantity in the first run. The heating rate was increased then over 40 min reaching 4.8 K/min and was kept constant for the remaining temperature range between 105 °C and 520 °C. Thus, the curves can be divided into three sections: Section 3.1 was the heating range from 25 °C and holding at 35 °C, Sect. 3.2 was the increasing heating rate up to 105 °C and Sect. 3.3 was the constant heating rate at 4.8 K/min. In the first section indeed the water concentration showed the most pronounced increase from the beginning at 25 °C, identified by the unified atomic mass unit (u) of 18. In Sect. 3.2 only the u values, which could be attributed to the atmospheric gases: 14 and 28 (N_2^{2+} and N_2^{+}) and 32 (O_2^{+}), as well as 40 (Ar^{+}), which was purged between the sample loadings, remained constant whereas the others increased significantly. Thus, only the gases which increased within this phase are supposed to have been emitted by the sample. These were: 18 (H_2O), 2 (H_2), 44 (may be CO_2 or acetaldehyde), 57 (maybe a 1:1 mixture of isobutylene and isobutane), 78 (maybe benzene) and with a very slight increase an unassigned compound with 64 u. As their release was mainly limited to the ramping Sect. 3.2 and did not lead to a significant decrease in the P2 bulk mass, it was supposed to indicate rather a desorption process of adsorbates or earlier damaged P2 molecules than the activation of a decomposition. Only two compounds did not reach a plateau in Sect. 3.3. These were those with the masses 44 and 78 u. Although these compounds, supposedly CO_2 or acetaldehyde and benzene may be decomposition products of P2, it is assumed, that they already existed within the aggregates, because their emission started from the beginning of the measurement and not after a distinct temperature threshold. Their continuing emission in Sect. 3.3, in contrast to the other expected adsorbates, may be explained by a slower desorption or diffusion rate out of the aggregates rather than by an activated decomposition. The first significant temperature activated change was the short water release at about 150 °C. This may be related to a de-complexation of a zinc–water complex at the surface of the aggregates, which therefore was not diffusion controlled and thus resulted in the very sharp peak. The next significant change arose at 220 °C, where the molecular mass, assigned to a mixture of isobutene and isobutane, got released peaking at about 250 °C. This is supposed to be the beginning of the decomposition by a release of the *tert*-butyl group.

The maximum detectable unified atomic mass unit was 250 u, because the system was not calibrated for heavier ions. Thus, it wasn't possible to detect whole molecules of P2 with 836 u. It is therefore not clear if P2 evaporated or if it decomposed entirely at lower temperatures. An indication for a decomposition was the rising of the ion mass 174 u, which had a similar curve progression at higher temperatures as 57 u and was correlated to a steep decrease of the P2 bulk mass. Additionally, the oxygen content decreased significant correlated to the bulk mass decrease and a significant fraction of the initial mass remained at 520 °C (9 %), which was supposed to be ZnO. That let us conclude, that P2 decomposed and did not evaporate completely. It can be concluded that P2 is stable up to about 220 °C.

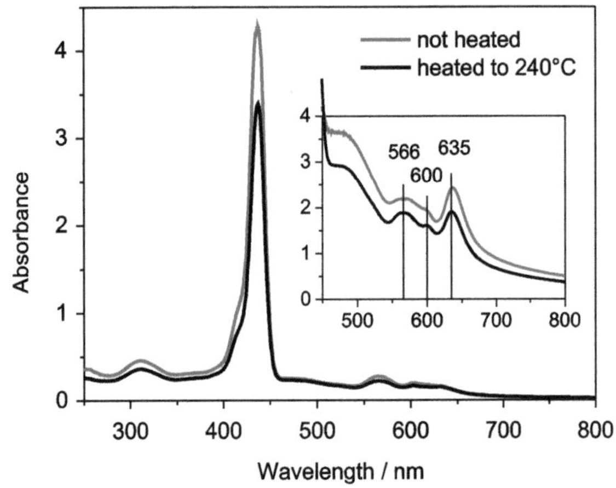

Fig. 3.20 Absorption spectra of P2 aggregates after a heating treatment up to 240 °C. A dry powder of P2 aggregates (J-aggregates) has been heated up to 240 °C (3 min between 230 and 240 °C) within air and dissolved afterward in commercial DCM (pro. anal.), (*black* lines). It is compared with a not heat treated reference solution of a roughly similar concentration (*gray* lines)

Another evidence for the stability of dry P2 aggregates even up to 240 °C was made by heating it for 3 min up to this temperature in a thin tube of approx. 1-mm diameter (open to air) and measuring afterward its absorption spectrum. For this purpose it has been dissolved within commercial DCM and the spectra, measured of different concentrations, showed no qualitative difference compared to the reference solution of untreated P2 (Fig. 3.20).

The inset shows the absorption spectra of higher concentrated solutions of the heated P2 and the untreated reference. Note that the absorption band at 635 nm is more intense at higher concentrations and that the spectra look slightly different compared to those measured in anhydrous DCM. The higher concentration leads to the formation of a higher fraction of J-aggregates, noticeable as increase of the band at 635 nm. The heat treatment was performed within the temperature controlled melting point analyzer *melting point apparatus SMP3* from *Stuart Scientific*. After a fast heating up to 230 °C the final heating period from 230 °C to 240 °C took about 3 min and no melting was observable.

The high thermal stability of the P2 aggregates was quite unexpected, as the 3-hydroxyethyl group could easily dehydrate thermally to a vinyl group as encountered in Chl *a*. This enhanced stability originates probably from the coordination of the zinc atom by the hydroxy group.

The morphology of P2 aggregates does also not change significantly upon annealing at 200 °C for 15 min in air, as could be seen in the SEM micrographs of a J-aggregate before and after the annealing (Fig. 3.21).

A similar annealing, but within N_2, was done in cooperation with T. Savenije and A. Huijser in Delft, which could improve the photosensitization of TiO_2 by P2 aggregates [11]. This treatment may have increased the adhesion to the TiO_2, which has been investigated afterward by SEM (Fig. 3.22).

Even though P2 is supposed to stick to TiO_2 surfaces by physisorption only, its J-aggregates appeared to have been relatively strong bound to such surfaces after the thermal treatment. The aggregates have been deposited by spin-coating from

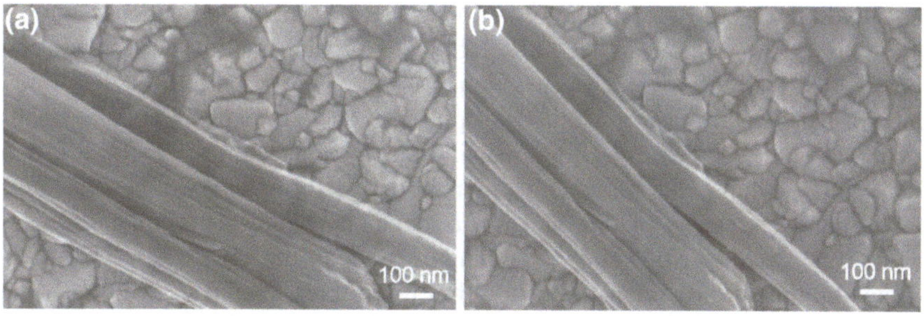

Fig. 3.21 SEM micrographs of a P2 aggregate before and after annealing at 200 °C. **a** P2 aggregate before annealing, as deposited by spin-coating from an anhydrous DCM dispersion onto an FTO-substrate (SnO_2:F). **b** The same P2 aggregate after annealing the substrate at 200 °C for 15 min in air. The substrate was slightly rotated after the second mounting into the SEM, but the aggregate did not diffuse significantly over the surface. In-lens, EHT: 20 kV, WD: 12 mm, aperture: 30 μm

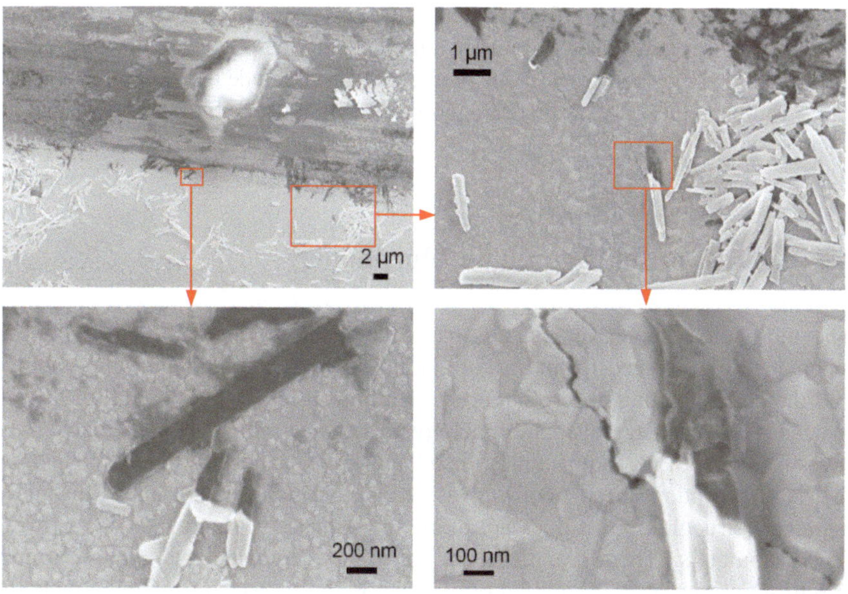

Fig. 3.22 SEM micrographs of P2 aggregates partly scratched off from TiO_2. P2 was spin-coated from anhydrous DCM onto a dense TiO_2 layer within an N_2-glovebox, and tempered at 200 °C for 15 min followed by a sputtering of the surface with 6-nm platinum. The surface has then been partly scratched, removing also some aggregates at the borderline. **a** The remaining porphyrin parts on the TiO_2 surface indicate a good contact to the substrate (**b–d**). The dark shadow in **d** around the residue of the partly removed aggregate, most pronounced on the right side of this residue, is most probably caused by shadowing the Pt sputtering by the aggregate. In-lens, EHT 20 kV, WD: 13 mm, aperture: 20 μm

an anhydrous DCM dispersion in order to measure the so-called incident photon to charge separation efficiency (IPCSE). This value is the fraction of incident photons, which lead to an electron injection from the aggregates into the TiO_2 conduction band upon the irradiation with pulsed monochromatic light. The IPCSE increased by a factor of 3-4 after annealing the aggregates on the TiO_2 surface at 200 °C for 15 min. One reason for this may be the formation of a better contact between the aggregates and the TiO_2 surface upon annealing. This contact area has been investigated by SEM after scratching off some of the aggregates. Prior to the scratching the surface was sputtered with a 6-nm thick Pt-film to avoid charging effects on the only 100 nm thin TiO_2 layer on the quartz substrate.

As can be seen most pronounced in Fig. 3.22d there is obviously some residue of the P2 aggregate sticking to the TiO_2 beneath the removed part of the aggregate. This part is thicker as the Pt-coating and thus is supposed to be a part of the P2 aggregate. The dark shadow around this residue, mainly on the right side of it, was probably caused by a shadowing effect of the aggregates during the Pt-sputtering. This shadow shows, that the contact area of this aggregate was not "glued" by the Pt layer. The residue, which could be seen under almost all removed P2 aggregates shows that the adhesion of P2 to the TiO_2 substrate was even stronger than the cohesion between the P2 molecules within the aggregate. Such a strong bonding to the substrate explains also, why many aggregates broke perpendicular to the long axes and did not get removed in one piece. We can conclude that a good contact on TiO_2, can be achieved with the P2 aggregates without specific binding groups for chemisorption.

3.7 Control of the Aggregate Morphology

The morphology of P2 aggregates has been investigated mainly by the SEM technique. We pursued to find a correlation between the aggregate morphology and the corresponding absorption spectra in order to have the possibility to estimate also the mean aggregate morphology on samples, which are not suitable to be investigated by SEM, for example, when the aggregates are embedded into another matrix. Absorption spectroscopy gives an average information about the mean aggregate structure, but it does not allow to estimate the aggregate size above a few nanometers. Therefore, the SEM technique complements the absorption spectroscopy very well, because it offers the possibility to characterize not only the average, but also the distribution of different morphologies beyond the size of a few nm. To find a correlation between the aggregate morphology and the absorption spectra it was necessary to control the morphology to achieve a narrow distribution of different aggregate morphologies. Therefor, different processing techniques have been employed to deposit aggregates as monodisperse as possible.

Another goal of this work was to achieve a more homogeneous, in the ideal case monodisperse, distribution of well-structured aggregates, which do not significantly exceed the exciton diffusion length at least in one dimension. That means that the typically obtained self-assemblies, from all investigated solutions so far, which do not hinder the self-assembly completely, need to be significantly reduced in size, while maintaining their ordered self-assembled structure to make them applicable within solar cells.

The first attempt we made to reduce the P2 aggregate size was to disrupt the dispersed aggregates within DCM by high-power ultrasonication with a titanium sonotrode. While increasing the sonication power the absorption spectrum of the cooled P2-DCM dispersion showed no change until the decomposition of the solvent occurred. The DCM released HCl which in turn decomposed the P2 by demetallation of the porphyrin, that is, dechelating the zinc atom. Mhuircheartaigh et al. observed similar effects with other chlorinated solvents, like chloroform, dichloroethane and dichlorobenzene, but he could not observe this effect for dichloromethane (DCM) [12]. As the sonotrode had a higher power, we were able to decompose also DCM, but this shows, that the aggregates are bound too strongly to be disrupted by this mechanical way. However, the aggregates may not exist within the DCM dispersion in that form in which they appear after deposition onto a substrate. These questions are addressed in the following sections.

3.7.1 Processing Techniques

Conventional dye-monolayer coatings for dye sensitized solar cells are being applied via a solution based chemisorption process. Thereby typically a meso-porous TiO_2 layer, sintered from a TiO_2 nanoparticle dispersion onto a substrate, is introduced for a few hours up to one day, depending on the temperature, into a dye solution and afterward rinsed with the blank solvent to wash off the residual, not chemisorbed dye. This technique is not applicable for P2, due to its lack of specific binding groups like a carboxylic group (COOH) and a rinsing with the same solvent would wash off the dye completely.

Such binding groups have not been attached to P2, because they may disturb its ability to self-assemble. Otherwise it may be possible, to coat a TiO_2 surface by a monolayer of a functionalized P2-derivative first, and than apply the original self-assembling P2, which then may grow epitaxially onto the first monolayer. In this thesis instead, it was investigated how self-assembling dyes without such specific binding groups can be controlled to sensitize efficiently a semiconductor surface.

The following methods for the deposition of P2 aggregates onto different substrates have been investigated: precipitation, dip-coating, spin-coating and spraying. These methods offered different possibilities to control the morphology of the coatings.

With the precipitation technique by mixing P2-DCM solutions/dispersions with *n*-heptane it was possible to achieve macroscopically homogeneous P2 aggregate coatings, but a control of the aggregate size and distribution on the micrometer scale was not possible. The dip-coating technique led to macroscopically very inhomogeneous films. The spin-coating technique offered the possibility to achieve macroscopically homogenous coatings, but the control of the micromorphology was limited. It was not possible to increase the dye-amount significantly by spin-coating the same solution several times onto the same substrate or dipping it several times into the same solution. The adsorbed dye amount could only be controlled by the concentration of the solution for this coating techniques. Thus, only highly concentrated solutions of P2 led to a sufficient high dye amount for the full sunlight absorption. High concentrations of P2 within DCM instead, lead to the formation of larger aggregates. Additionally, a lot material is wasted or at least used up and impurified by this technique in combination with concentrated solutions.

The spraying or airbrushing technique offers many possibilities to control the micromorphology, while allowing a good macroscopic homogeneity. It allows the repetition of many coating steps from diluted solutions to control the micromorphology and achieve every desired thickness of the coating. This efficient coating technique is also easy to upscale onto large areas or implement in a roll-to-roll process. However, it is difficult to reproduce the coatings exactly with a manually controlled spraying pistol, because many parameters influence the film formation, like the duration of the spraying pulses, temperature and the solvent vapor pressure in the atmosphere. These parameters influence the evaporation of the droplets and thus their size at the moment of their impact onto the sample. Additionally, with the manual control of the pistol a variation of the spraying pulse duration was unavoidable, thus a longer pulse could cause an increased fusion of the droplets on the sample surface, which had a pronounced effect on the micromorphology, meaning not only the morphology of single aggregates but also their distribution in the micrometer scale. To use the full potential of this technique an automated spraying system would be required.

Another possibility to control the aggregate size-distribution is the usage of a structured substrate as template for the aggregate growth. In an earlier work, we demonstrated the possibility to infiltrate P2 aggregates into mesoporous TiO_2 layers, which contained additional macropores, in which the aggregates grew with their size confined by the diameter of the macropores [13, 14]. The macropores have been introduced into the mesoporous TiO_2 layers by mixing polystyrene spheres into the TiO_2 particle dispersion, which has been applied onto a substrate in form of thin layers by doctor-blading. During the sintering process the polystyrene was pyrolyzed, leaving behind well-defined and homogeneously distributed macropores. As the aggregates could not be infiltrated through the mesopores directly from DCM, a two step process with different solvents was developed. In the first step P2 was dissolved completely to monomers within THF, from which it could be infiltrated into the porous layer. In the second step the resulting aggregates, which were apparently more disordered in structure, if not completely amorphous, could then be reorganized by a thermal post-treatment within *n*-heptane, a solvent, which facilitates the self-assembly without dissolving P2. This technique has been further investigated and improved in this work.

3.7.2 *P2 Infiltration into Mesoporous TiO_2-Structures*

The basic idea was to infiltrate mesoporous TiO_2 layers, which contained additional macropores, with P2 from monomeric solutions in THF and induce a reorganization process of the amorphous phase to achieve a J-aggregate growth within the macropores. As it was not possible to infiltrate such layers from anhydrous P2-DCM solutions, even at 110 °C under 5 bar overpressure in a pressure tube, it was assumed that, if it is once infiltrated, it could then not easily re-diffuse outwards. This would allow to treat the P2-deposits with various solvents at different temperatures to optimize the aggregate morphology for a proper function in such modified dye sensitized solar cells. The prerequisite therefore would be that the aggregates segregate out of the mesopores and tend to grow into the free space of the macropores.

It turned out, that the macropores could be filled only then significantly with P2, when high concentrations of P2 in THF were used. To investigate the infiltration of P2 at different concentrations the samples were infiltrated intentionally inhomogeneous from THF by the drop-casting technique. This resulted in many different pore-filing ratios and concentration gradients on one sample, which could be compared.

Figure 3.23 shows a mesoporous TiO_2 layer on a glass substrate, equipped with additional 500-nm-sized macropores, which was infiltrated with P2 by drop-casting from a THF solution and letting it dry in the air.

In the overview (inset in a), it can be seen that the coating is thicker at the edge, and forms some dark lines of a P2 overlayer, which peels off at the edge of the sample. The straight vertical line is a scratch in the layer, which has been introduced after the infiltration to have a view inside the inner pores. The rectangle marks the area magnified in (a), which is relatively homogenously coated with an intermediate concentration of P2, as can be estimated from the grayscale in the overview. The bottom in (a) is more bright, thus less covered by P2. b-d give an inside view into the inner macropores, with many of them containing a spherical cluster of P2. In other areas of the sample with different P2 concentrations the inner macropores remained empty. In areas with a lower P2 concentration no aggregates could be distinguished from the mesoporous matrix, they most probably resided only within the mesopores. In areas with a higher P2 concentration the overlayer tended to peel off at a certain thickness, as can be seen on a similar sample in Fig. 3.24.

3.7.3 n-*Heptane Induced Reorganization Within Mesoporous TiO_2*

The apparently amorphous deposits from THF were treated with *n*-heptane in order to attain size-limited J-aggregates within the macropores. Thereby the infiltrated samples were placed within a pressure-tube of glass and immersed in anhydrous *n*-

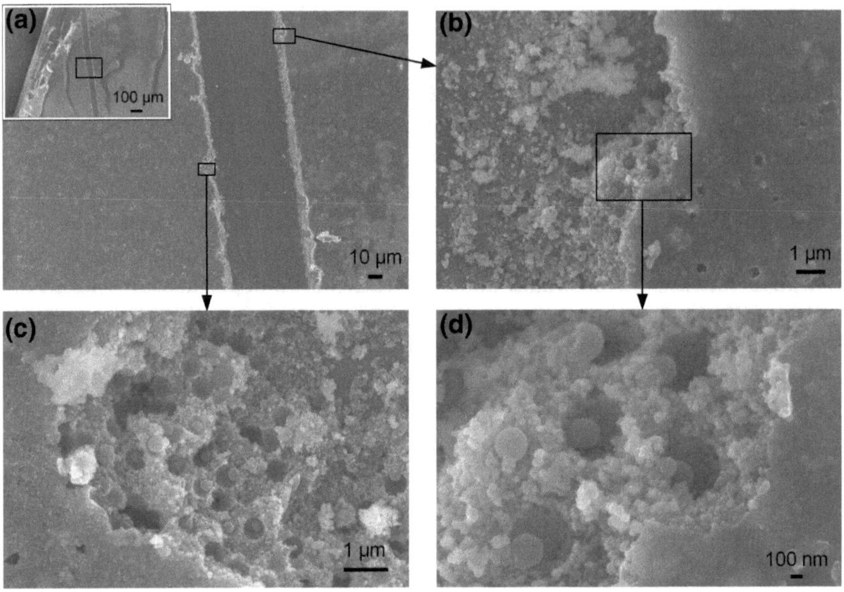

Fig. 3.23 SEM micrographs of P2 drop-cast from THF onto mesoporous TiO_2. One drop of a 1 mM solution of P2 in anhydrous THF was applied onto a mesoporous TiO_2 layer, which contained additional 0.5 µm-sized macropores. It was left dry in air. The inset in (**a**) shows an overview of the coated substrate with its edge on the left and a vertical scratch, which was made after the infiltration. Note the inhomogeneous infiltration and the dark P2-overlayer, which peels off at the edge of the substrate. The rectangles mark the areas magnified in (**b–d**). As the substrate was glass, the sample was sputtered with a 5-nm thick Pt layer. In-lens, EHT: 10 kV, WD: 5 mm, aperture: 30 µm

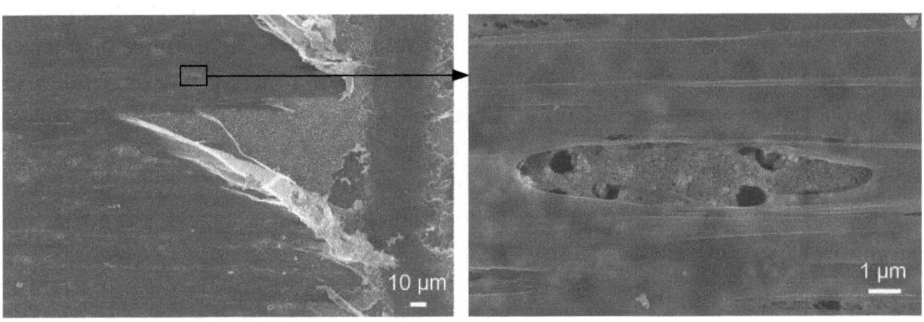

Fig. 3.24 SEM micrograph of P2 deposited by drop-casting from THF onto mesoporous TiO_2. Approx. 10 drops of a P2-THF solution have been dried onto a mesoporous TiO_2 layer with 1 µm macropores. The formed P2 layer peeled off in certain areas. The surface was sputtered afterward with 2.4 nm Pt. In-lens, EHT: 10 kV, WD: 6 mm, aperture: 30 µm

heptane. The empty space of the tube was filled with N_2. Subsequently the tube was closed tightly and annealed within an oven at 110 °C up to 4 h.

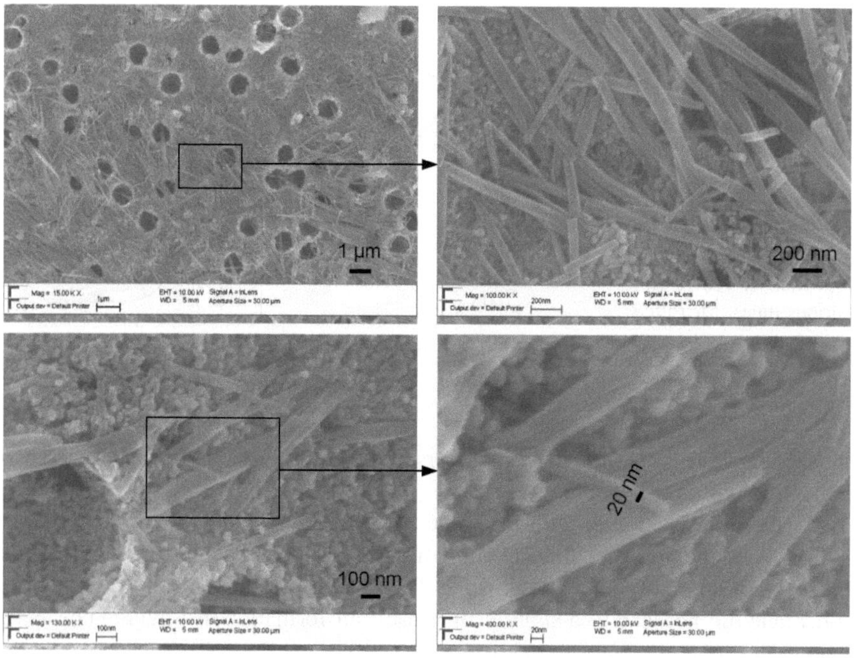

Fig. 3.25 SEM micrographs of P2 aggregates onto mesop. TiO_2 after annealing in *n*-heptane. The same P2-THF solution as used for the sample in Fig. 3.24 was drop-cast onto this mesoporous TiO_2 layer with additional macropores. The sample was then annealed within 3 ml of anhydrous *n*-heptane in a pressure-tube at 110 °C for 3 h and 40 min. The surface was sputtered with a 5.2-nm thick Pt layer

After short annealing times within *n*-heptane in the order of tens of minutes, small elongated aggregates could be seen within the inner macropores in areas with a high initial pore-filling with amorphous P2 deposits. After longer annealing times however, the aggregates within the inner macropores did not grow significantly, actually most of them disappeared, while larger aggregates grew on top of the TiO_2 layer as can be seen in Fig. 3.25.

Whereas a significant aggregate growth could be achieved within the inner macropores after prolonged annealing times within *n*-heptane, the aggregates, which grew on top of the TiO_2 layers reached a high aspect ratio with a very thin diameter. Thus, these aggregates do not necessarily need to be confined in their length to achieve a small diameter.

For investigations by absorption spectroscopy the prepared samples needed to be macroscopically more homogenous to allow an interpretation of the averaged signal over the roughly 0.5 cm^2 big irradiation spot. To deposit the high concentration of P2, which was necessary to obtain aggregates within the macropores, homogeneously over this large area, the P2-THF solution was applied by the spraying technique. The absorption spectroscopy in Fig. 3.26 revealed that the

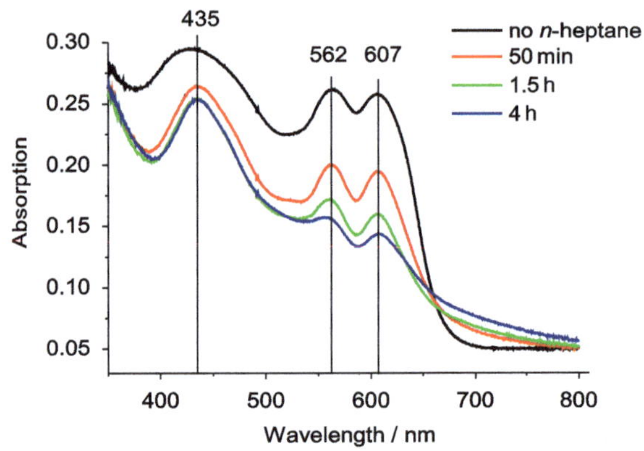

Fig. 3.26 Absorption spectra of P2 within mesoporous TiO_2 + macropores, annealed in *n*–heptane. A 0.5-mM solution of P2 in anhydrous THF has been sprayed onto a mesoporous TiO_2 layer with 1-μm-sized macropores followed by annealing within anhydrous *n*-heptane at 110 °C for various times. Measured within an integrating sphere

mean aggregate structure in such layers remains amorphous, even after 4 h of annealing within *n*-heptane.

As these absorption spectra resemble those of P2 monomers within THF with similar band maxima and almost symmetrically broadened bands, they indicate the presence of an amorphous phase of P2. The increase of the annealing time within *n*-heptane did not lead to a significant phase-transformation to J-aggregates as the band did not significantly red-shift.

It can be concluded that the aggregates do not really tend to grow into the macropores upon the post-treatment in *n*-heptane. They seemingly only reorganize there from amorphous P2-clusters, segregated within the macropores in advance. The major fraction of P2 instead, remained most probably within the mesopores, invisible for the SEM but detectable by absorption spectroscopy.

The thin aggregates, which have been seen on top of the TiO_2 layers indicate, that the aggregates do not necessarily need to be confined in their length to achieve a small diameter, because obviously their aspect ratio can also be increased preserving a small diameter. As a similar thin substructure has been found also in precipitated aggregates, which have been shown in Fig. 3.13b, it has been investigated in the following, if the aggregates initially grow to such thin rod-shaped structures and only agglomerate later on to the undesired thick ones.

3.7.4 Precipitation Between ZnO Nanorods

It has been investigated whether any agglomeration takes place during the aggregate deposition and if it could be perturbed by a suitable design of the substrate. Therefore, it has been chosen a substrate with vertically aligned ZnO nanorods with lengths around 1 μm. These substrates were prepared by the group of Heinz Kalt at KIT, where Jonas Conradt tested them as electrode materials for solar cells.

Fig. 3.27 SEM micrographs of P2 as precipitated from *n*-heptane onto ZnO nanorods. These micrographs show agglomerates of P2 J-aggregates on a substrate with vertically aligned ZnO nanorods. The nanorods hindered partly the aggregates to agglomerate. The preparation was done within a glove box by adding an anhydrous P2-DCM solution into a major amount of *n*-heptane in which this substrate was placed. The rough structure at the bottom belongs to the ZnO nanorod substrate, and not to the P2 aggregates. Sputtered with 5.2 nm Pt

The deposition of P2 aggregates, by the precipitation technique from DCM in *n*-heptane, occurs predominantly onto solid surfaces, which are immersed by the solution. As the aggregates grow also on surfaces directed downward, it was expected that the agglomeration process takes place predominantly at the substrate surface and not only within the dispersion with subsequent falling onto the surface, as this could happen only when the investigated surface would have been directed upward. The precipitation of P2 was induced onto the ZnO nanorod substrates by injecting an anhydrous P2-DCM dispersion into anhydrous *n*-heptane in which the substrates were placed directed upward, due to their small size and hence difficult handling. The results are shown in Fig. 3.27.

As can be seen, indeed an agglomeration effect takes place at or close to the surface. Many examples could be seen, where the ZnO nanorods were embedded within the agglomerates, like in the thick one visible in the top left corner of the large micrograph. In this agglomerate, a thin aggregate shows a strong curvature bending around the ZnO nanorod. Some thicker primary aggregates instead, did not bend so far, and thus were partly hindered in their agglomeration, as can be seen also in the small micrographs.

Although it is not entirely clear which of these rods belong to the P2 aggregates and which may be broken ZnO nanorods, it is obvious that an agglomeration process of thinner P2 aggregates to thicker ones takes place during the deposition.

3.7.5 *Drop-casting from DCM Dispersions at Low Temperatures*

If the aggregates agglomerate predominantly during the deposition process, it might be possible to slow down or completely inhibit this process by cooling the dispersion during the deposition. This has been tested by drop-casting P2 from a cooled anhydrous DCM dispersion under vacuum. Therefore, the dispersion was dropped onto an FTO substrate at −108 °C under dry N_2 atmosphere. It froze immediately – the melting point of DCM is at about −97 °C. This chamber was then evacuated to about 0.04 mbar in order to enhance the DCM sublimation. As the solvent did not sublime after several hours the temperature was allowed to increase. At −60 °C the dispersion finally melted and evaporated in the vacuum completely within a few min at a temperature still below −50 °C. The resulting P2 deposits on FTO are shown in Fig. 3.28.

These structures resembled clews of yarn. The aggregates looked like thin crumpled fibers in which a few bigger conventional rod-shaped aggregates were embedded. In contrast to these typical big rod-shaped aggregates, the P2-fibers were very sensitive to the electron beam of the SEM. They seemed to melt within seconds at the electron irradiation doses of the magnifications shown here. These micrographs may represent the morphology of the aggregates within dispersions in DCM, as the agglomeration process may have been completely suppressed at this temperature. The fiber-shape could be the reason, why it was not possible to infiltrate the mesoporous TiO_2 layers from DCM dispersions even at 110 °C (under pressure).

Figure 3.29 shows the absorption spectrum of this sample together with that of a reference sample, which has been drop-cast from the same dispersion at room temperature.

Despite the non-ideal conditions for absorption measurements on the not fully transparent FTO substrates, a significant difference in the absorption spectrum of the P2 aggregates, deposited at about −50 °C and those at room temperature could

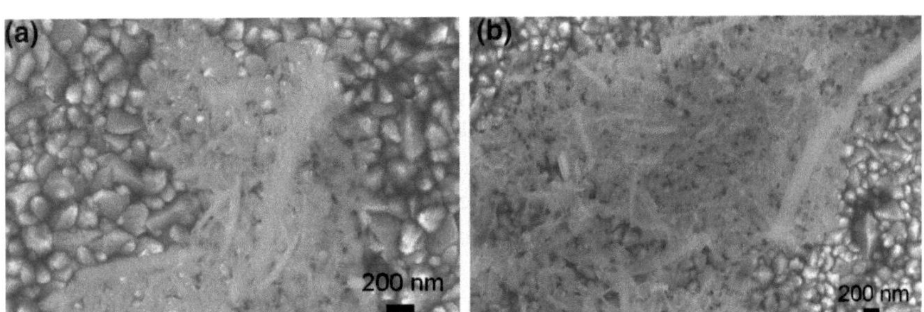

Fig. 3.28 SEM micrographs of P2 deposited onto FTO from a cooled DCM solution. These fiber-like P2 aggregates on an FTO substrate were deposited by drop-casting from a P2-DCM dispersion at about −50 °C under vacuum (0.04 mbar) after an initial freezing at −108 °C. In-lens, EHT: 5 kV, WD: 5 mm, aperture: 20 μm

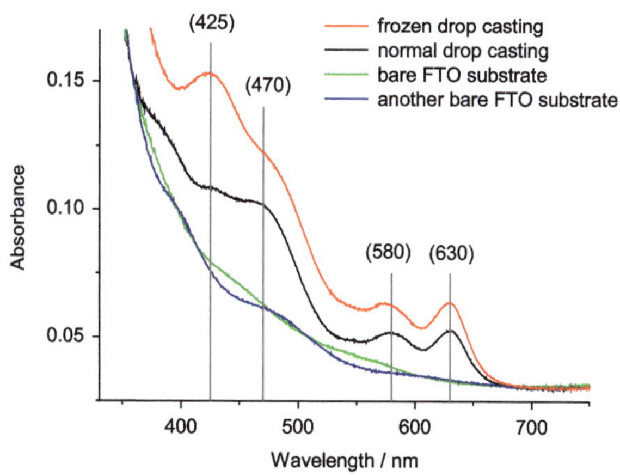

Fig. 3.29 Absorption spectrum of P2 on FTO, deposited from a cooled DCM solution. A few drops of a P2-DCM dispersion were frozen onto an FTO substrate at −108 °C. Under a vacuum of $4 \cdot 10^{-2}$ mbar the solvent has been left to warm up. At about −60 °C the DCM dispersion melted and evaporated within a few minutes in the vacuum at a temperature still below −50 °C. The deposits of P2 on FTO (*red*) are shown in comparison with a reference sample where a few drops of the same P2-DCM solution have been dried at ambient conditions at 1 bar and 21 °C in air (dark). The *green* and *blue* curves show the absorption of typical bare FTO substrates showing the small variations in their absorbance, which may slightly influenced the absorption spectra of the P2 deposits. Measured within an integrating sphere

be noticed, anyway. Although the Q band region was similar of both samples, the higher Soret band absorbance of the aggregates deposited at about −50 °C might indicate a higher fraction of disordered aggregates. It could result from an amorphous P2 residue or also from the fiber-shaped aggregates themselves, which had, due to their flexibility, a very short range order, where the bended regions could have a very distorted structure. As the Q-bands were red-shifted like in conventional J-aggregates, it can assumed that the fiber-like aggregates had a J-aggregate structure. The absorption spectrum of the aggregates deposited under cooled conditions resembles more that of the liquid dispersion, due to the higher Soret band. Hence, it can be concluded that a significant agglomeration to bigger aggregates indeed takes place during the short evaporation time at room temperature and that this could be slowed down by lowering the temperature.

This was the first time, that such a morphology of P2 aggregates has been observed. These fibers probably agglomerate by sticking together within a few seconds, during which the highly volatile solvent evaporates. Thus, the deposition temperature is a very important parameter to influence the morphology of the aggregates and agglomerates. It can be also inferred from this, that it might be possible to hinder the agglomeration process also by the spray deposition technique, as this limits the amount of P2 within the small droplets and accelerates the solvent evaporation. This might hinder the aggregate growth, when the sprayed droplets do not merge on the substrate before the solvent evaporates.

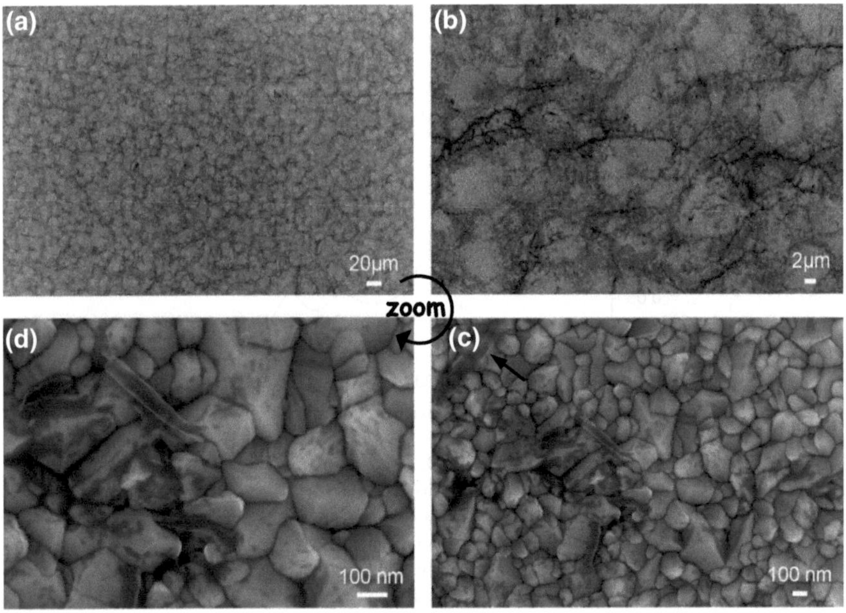

Fig. 3.30 SEM micrographs of P2 sprayed from a DCM dispersion onto FTO. This magnification series shows a representative morphology of P2 as sprayed from a dispersion in anhydrous DCM on an initially perfect clean FTO substrate. The arrow in (**c**) marks a larger, crystalline appearing, aggregate, beside mostly amorphous appearing spots of P2. In-lens, EHT: 10 kV, WD: 5 mm, aperture: 30 μm

3.7.6 Spraying P2 Aggregates from DCM

As a conclusion of the previous experiments it can be assumed, that the spraying technique has the potential to reduce the size of the deposited aggregates. A manually controlled spraying pistol was mounted at defined distances and angles relative to the substrate and dry nitrogen was used as carrier gas at 1.5 bar. The P2-DCM dispersions have been sprayed in short pulses in order to minimize the cooling effect on the sample. This might have caused a reduced evaporation rate and hence, merging of the droplets onto the substrate allowing the formation of larger agglomerates. Figure 3.30 shows a representative sample of the sprayed P2 aggregates on an FTO substrate.

The mean aggregate size could indeed have been reduced, but the overall result is not satisfying. The deposited P2 aggregates lost much of their crystalline appearance as can be seen in (d). Whereas macroscopically the coating looks very homogenous (a), on a microscopic level instead, the distribution of the aggregates is very inhomogeneous (b-d). The mean aggregate shape was resembling that of amorphous spots rather than highly ordered J-aggregates. Between these spots also some bigger aggregates could be found, which appeared more crystalline, due to their smooth facets and higher transparency to electrons, as that one marked by the arrow in (c).

The absorption spectrum of this sample looked qualitatively very similar to that of the fiber-like deposits from the cooled P2-DCM dispersion, which was shown in Fig. 3.29 and quantitatively the absorbance was half that of the fiber-like aggregate sample. All variations of the spraying parameters did not lead to a significant improvement of the coating's homogeneity.

I can be concluded, that although the mean aggregate size could be reduced significantly by the spraying technique the assumed loss in structural quality would require a post-treatment. This may be the annealing within *n*-heptane, which proved of value already for the reorganization of the amorphous deposits from THF-solutions. The main advantage of the processing from DCM dispersions was the possibility to directly obtain well-defined J-aggregates. As this advantage was lost by the attempt to reduce the aggregate size, there is no need to work with DCM dispersions anymore, when a two step process is necessary, anyway. From THF solutions it is at least not problematic to deposit very homogenous films, for example, by spin-coating, due to their amorphous nature. Thus, we went on investigating this two step process using smooth substrates.

3.7.7 Reorganization of Amorphous P2 to J-Aggregates on Flat Substrates

Thin films of amorphous P2 have been deposited by spin-coating from THF-solutions onto glass substrates and subsequent annealed within anhydrous *n*-heptane in order to induce a reorganization to J-aggregates. As during the annealing within *n*-heptane sometimes parts of the film were washed off, it was tried if the *n*-heptane may also catalyze the reorganization process from the vapor phase. Whereas the annealing within air, N_2 or vacuum had no significant effect on the morphology of the films, the presence of *n*-heptane vapor in an annealed pressure tube had indeed a similar effect as when the films were immersed within the *n*-heptane.

A film of P2 has been deposited onto glass by spin-coating from a 6 mM solution in THF. The annealing at 110 °C for 4 h in *n*-heptane vapor caused the film structure to reorganize to similar rod-shaped aggregates, as known from their direct growth in DCM, but with a smaller size and a significant improvement in the homogeneity and size-distribution, as can be seen in Fig. 3.31.

This is an impressive improvement of the homogeneity of the aggregate distribution, locally and in size. Additionally, only a small fraction of the aggregates is piled one over another, what makes this structure already suitable for solar cells. Due to the insulating glass substrate, a sputtered gold coating with 6-nm thickness was needed to avoid charging effects. The gold film was not smooth and caused a rough substructure in the order of a few tens of nm, but the aggregates beneath it

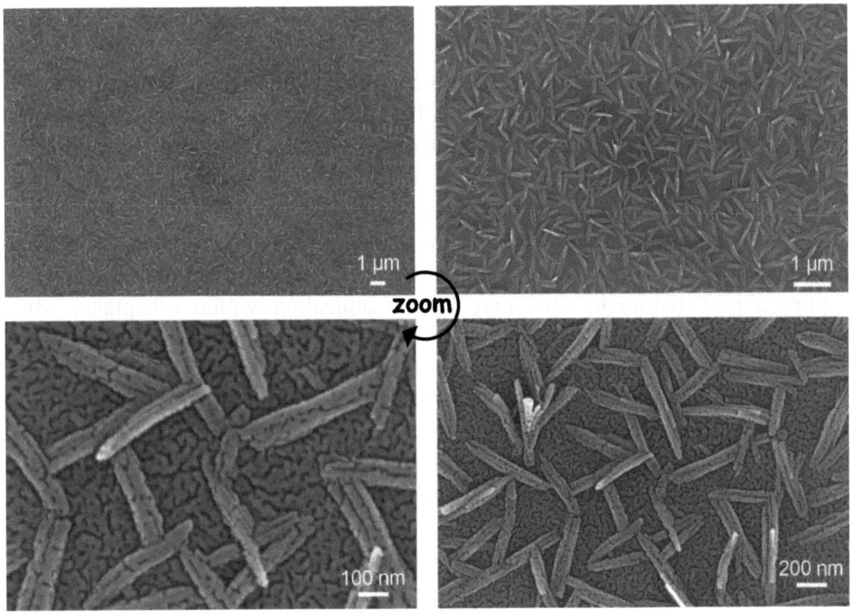

Fig. 3.31 SEM micrographs of P2 spin-coated from THF onto glass and annealed in *n*–heptane vapor. A 6 mM solution of P2 in THF was spin-coated at 3000 revolutions per minute onto a glass substrate, which has been afterward annealed at 110 °C in *n*-heptane vapor for 4 h. The sample was placed within a pressure tube containing a separated reservoir with *n*-heptane, which was not wetting the sample, but only affected it through the gas phase. Note the relative monodisperse aggregate size with very homogeneous distribution over the entire surface. The substructure originates from the sputtered 6-nm thick gold layer. In-lens, EHT: 10 kV, WD: 5 mm, aperture: 30 µm (**b**, **d**). Reprinted with permission from Chappaz-Gillot C, Marek PL, Blaive BJ, Canard G, Bürck J, Garab G, Hahn H, Jávorfi T, Kelemen L, Krupke R, Mössinger D, Ormos P, Reddy CM, Roussel C, Steinbach G, Szabó M, Ulrich AS, Vanthuyne N, Vijayaraghavan A, Zupcanova A, Balaban TS (2012) J Am Chem Soc 134:944-954. Copyright 2012 American Chemical Society

can be clearly recognized. The glass was used to allow a better characterization by absorption spectroscopy, which is shown in Fig. 3.32.

The absorption spectrum of the dry P2-film, as grown from the THF solution, resembles very much the absorption spectrum of P2 monomers within THF solutions, with the difference, that the spectrum of the dry film is slightly broader. This proves, that the film was amorphous, as has been also expected from the flat films seen by the SEM. After the annealing within *n*-heptane, the absorption spectrum transformed into the red-shifted one, which is known from J-aggregates grown directly from DCM solutions, as shown in the comparison. The red-shifted Soret band of the reorganized J-aggregates at 455 nm is not a shoulder anymore, but developed to a distinct band with an even higher absorbance than the original Soret band at 431 nm. This shows that either the structure of the J-aggregates

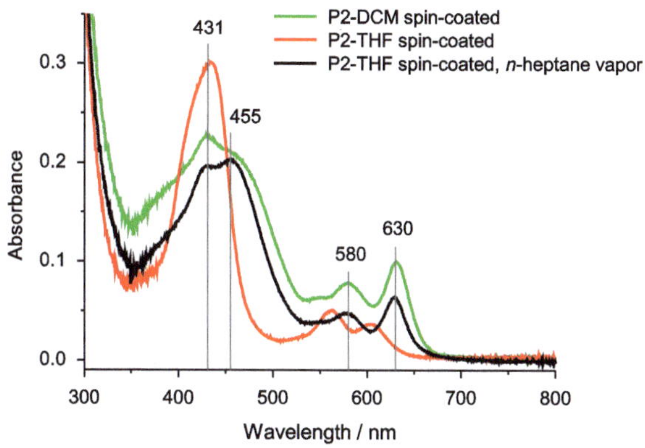

Fig. 3.32 Absorption spectra of P2 from THF on glass with annealing in *n*-heptane vapor. These absorption spectra were measured of the film as spin-coated from the 6 mM P2-THF solution on glass (*red*) and after the thermal treatment within *n*-heptane vapor at 110 °C for 4 h, before the sputtering with gold (*black*). For comparison also a glass substrate with P2 is shown as spin-coated form a 6 mM P2 dispersion in anhydrous DCM (*green*). The spectra were recorded within an integrating sphere

improved by a decrease of the defect-density or the quantity of a possible secondary amorphous phase decreased.

Thus, together with the improved dispersity, these J-aggregates, reorganized from amorphous films, are significantly better suited for solar cells, then those directly grown from solvents which facilitate the self-assembly.

3.7.8 P2 Reorganization Within Porous TiO_2 Layers

To increase the absorbance and to further reduce the aggregate size, especially the diameter, which may still exceed the exciton diffusion length, the same technique was applied to infiltrate coarse-porous TiO_2 layers with P2 J-aggregates.

In general an increase of absorption while maintaining a short exciton path way through the aggregates to the TiO_2 interface would allow to increase the external quantum efficiency (EQE). The EQE is the ratio of the incident photons to electrons within the external current. It is measured wavelength-resolved and also called: incident photon-to-current conversion efficiency (IPCE). For the internal quantum efficiency (IQE) instead, also called absorbed photon-to-current conversion efficiency (APCE), only the absorbed photos are regarded for this ratio. Thus, a thin layer, which absorbs only a fraction of the incident light may have an IQE of 100 % but a lower EQE. Important is therefore the EQE, as it shows if this system is already applicable for efficient solar cells under high light intensities.

Three different TiO_2-particle dispersions with different particle sizes of the anatase modification were used to sinter TiO_2 layers with different pore-sizes.

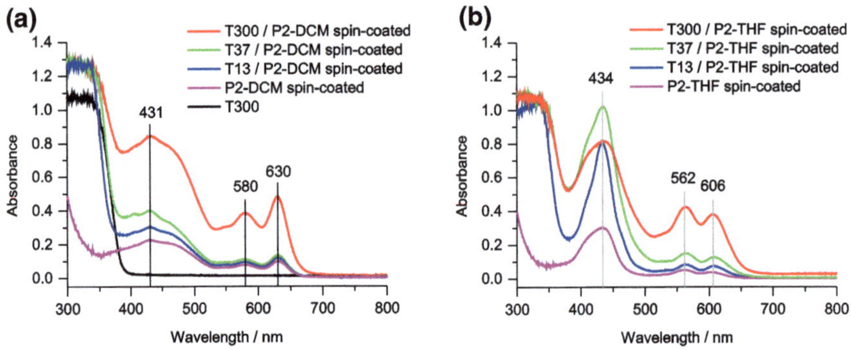

Fig. 3.33 Absorption Spectra of P2 spin-coated from DCM and THF onto different porous TiO_2 layers. T13, T37 and T300 indicate TiO_2 layers on glass substrates, sintered from TiO_2-particles with diameters up to 13, 37 and 100–300 nm, respectively. Substrates without any TiO_2 layer were bare glass panes. P2-DCM: onto each sample have been spin-coated 10 μl of a 6 mM dispersion of P2 within anhydrous DCM, P2-THF indicates the spin-coating of 10 μl of a 6 mM solution of P2 within THF. For comparison is also shown a bare TiO_2 layer T300 without P2 (**a**). The spin-coatings were performed at 3000 revolutions per minute (rpm). The measurements have been done within an integrating sphere

These dispersions, purchased from *Solaronix SA*, were readily prepared to be used by the doctor-blade technique and contained also organic binders, which prevented the packing to be too dense. Thus, highly porous, homogenous and crack-free layers were obtained. They were named T13, T37, and T300 as they contained anatase particles with sizes up to 13, 37, and 300 nm in diameter, respectively. The dispersion T300 contained also a small fraction of anatase-nanoparticles around 10 nm in diameter, which served as sintering aid to reduce the sintering temperature. From SEM micrographs the mean particle diameter of the T300 was estimated to be within the range of 100-300 nm in average, beside the small fraction of nanoparticles. No additional macropores were introduced into these layers, thus the resulting mean pore-sizes were in the order of the mean particle sizes. The layers were doctor-bladed onto glass substrates and sintered at 450–500 °C for 1–1.5 h and had a similar thicknesses around 4 μm.

Figure 3.33 shows the absorption spectra of the three different porous TiO_2 layers onto which P2 has been spin-coated with the same concentration and preparation parameters, but from two different solvents: anhydrous DCM and THF.

As can be seen in Fig. 3.33a, the absorption spectra of P2 differ only quantitatively within/onto the differently porous TiO_2 layers. The absorbance of P2 increases with increase of the TiO_2-particle size probably due to the higher light scattering of particles when their size approaches the irradiation wavelength. The amount of P2 obviously did not increase significantly from the bare glass substrate to the two TiO_2 layers of 13 and 37 nm particles, as can be seen in the similar height of the Q-bands. This agrees with the observations that P2 can not be infiltrated into mesopores from DCM, thus the aggregates are deposited only on top of these layers. The absorbance increased mainly at small wavelengths in the range of the Soret band, due to the increased scattering effect in this wavelength range. The TiO_2 layer with the particle size and hence pore-diameter around 100–300 nm probably has been penetrated partly by the aggregates from the dispersion in DCM. The scattering effect is very beneficial to increase additionally the light-harvesting. The light scattering effect of the porous TiO_2 layers may be better estimated on the P2-infiltration series from THF solutions (b). These solutions penetrated all of the differently porous layers, what could be seen also by the naked eye, because the layers looked from the backside through the glass substrates similar intense green as from the front side, in contrast to the samples spin-coated from P2-DCM dispersions, which looked much more pale from the backside than the front side. The improved absorbance toward bigger TiO_2-particles can be related to the increased light scattering. Qualitatively the absorption spectra of the infiltrated mesopores differed slightly from those, with the coarse or no pores. Whereas all spectra revealed no J-aggregate structure, as there were no red-shifted Q-bands, the spectra of P2 on bare glass and within the T300 layer had a relatively small Soret band. Hence, the absorption spectra of the mesoporous TiO_2 layers (T13, T37) infiltrated with P2 from THF solutions resembled more that of the liquid solution in contrast to the P2 deposits, which were less confined by the pore-size. This might be an indication, that the strongly size-confined deposits within the mesopores are more disturbed in their intermolecular ordering than those, which had more space to grow. As there is almost no qualitative difference between the absorption spectra of P2 within the T300 layer and P2 on the flat glass substrate, it can be concluded, that the pores of the T300 layer were large enough to allow the same degree of intermolecular ordering as on a flat surface, although THF hampers the self-assembly, anyway.

All these samples have been treated simultaneously in a single experiment within the same pressure tube and *n*-heptane vapor at 110 °C for 4 h in order to study only the influence of the pore-size on the reorganization process and exclude any other effects, which might have occurred in separated experiments, like small variations in time, temperature and humidity.

Although SEM-micrographs of mesoporous TiO_2 layers infiltrated with amorphous P2-deposits revealed after a thermal treatment in *n*-heptane or *n*-heptane vapor the presence of rod-shaped aggregates, absorption spectroscopy measurements showed that the major fraction of P2 remained amorphous. This amorphous phase is probably located within the mesopores, thus nearly impossible to see and characterize by SEM. The absorption spectroscopy allowed to detect the mean aggregate structure, that is, the structure of the major fraction of P2, and showed that this remains amorphous in the presence of mesopores with sizes up to 37 nm. Furthermore, not only the reorganization of the amorphous phase to J-aggregates was hindered by the mesopores, but also previously existing J-aggregates, which have been spin-coated directly from anhydrous DCM on top of these layers, disappeared upon annealing in *n*-heptane vapor, what could be proved also by absorption spectroscopy, as shown in Fig. 3.34.

As can be seen in Fig. 3.34 the spectrum of amorphous P2 within the TiO_2 layer with the maximum pore-size of 37 nm, could not be reorganized to gain the

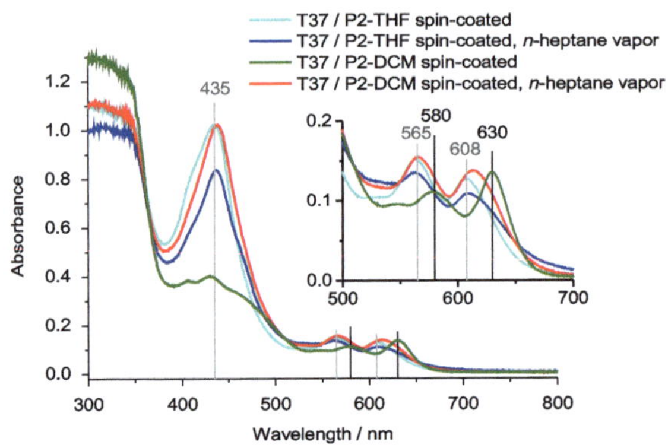

Fig. 3.34 Absorption spectra of P2 within mesoporous TiO_2 layers + *n*-heptane vapor treatment. Both TiO_2 layers from 37 nm particles, which were coated by P2 from anhydrous DCM and THF have been annealed together in the same batch at 110 °C for 4 h within *n*-heptane vapor in N_2. The absorption spectra of these layers are shown before (light *blue* and dark *green*) and after the annealing (dark *blue* and *red*), respectively. The inset shows a magnification of the Q-band region. Marek PL, Hahn H, Balaban TS (2011) Energy Environ Sci 4:2366–2378 – Reproduced by permission of The Royal Society of Chemistry

absorption spectrum of J-aggregates upon the annealing in *n*-heptane-vapor. A surprising result was that P2, which has been deposited on top of the TiO_2 layer from anhydrous DCM in form of J-aggregates changed its organization toward a more amorphous one. Such behavior could not be observed on flat surfaces and on the T300 layer. Most probably the original J-aggregates disassembled at the interface to the mesoporous TiO_2 and P2 diffused into the mesopores and remained there in the amorphous phase. This agrees with the previous observations, that P2 J-aggregates could not be found within macropores embedded in a mesoporous matrix unless the concentration of P2 was high enough to fill most of the pores.

The reorganization of the amorphous P2 into the J-aggregate structure was possible in the coarse-porous TiO_2 layer T300, instead. The absorption spectrum in Fig. 3.35 shows the gain in absorbance of reorganized P2 J-aggregates within the macropores of T300, compared to those on the previously shown flat substrate.

As can be seen the mean P2 aggregate structure could be transformed from the amorphous state into that of J-aggregates within the coarse-porous TiO_2 layer similar, as on the flat substrate. The absorbance increased more than three times using the coarse-porous substrate what can be attributed to the increased light scattering and a potentially higher dye uptake from the same solution.

The corresponding SEM micrographs of the T300 sample with the reorganized P2 J-aggregates and a blank T300 layer are shown in Fig. 3.36.

As can be seen the J-aggregates could be additionally reduced in size by the confinement within these pores. Due to their relative high aspect ratio around 7, their mean diameter within 200-nm sized pores, still remains below 30 nm. Thus, for aggregates, which stick with one side to a TiO_2 particle within a 200-nm sized pore an exciton diffusion length of 30 nm would be sufficient to convert almost every

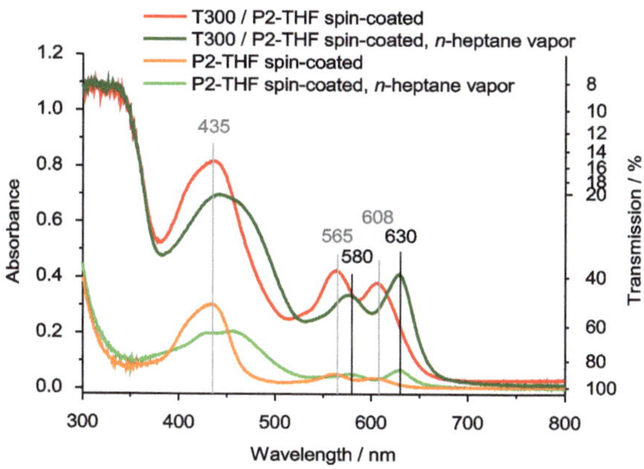

Fig. 3.35 Absorption spectra of P2 reorganized in *n*-heptane vapor within coarse-porous TiO_2. P2 has been spin-coated from THF onto a flat glass substrate (*orange*) and onto the coarse-porous TiO_2 layer T300 on glass (*red*). Both samples have then been annealed simultaneously in the same *n*-heptane vapor at 110 °C for 4 h in a pressure tube (light *green* and dark *green*, respectively). 10 μl of a 6 mM solution of P2-THF have been spin-coated at 3000 rpm on each sample. The absorption spectra were recorded within an integrating sphere. Marek PL, Hahn H, Balaban TS (2011) Energy Environ Sci 4:2366–2378 – Reproduced by permission of The Royal Society of Chemistry

absorbed photon into separated charges at the TiO2 interface injecting electrons into its conduction band. If the rest of the J-aggregate surface could be covered by a hole transport material (HTM) forming an interface at which excitons might also split, this time by injecting holes into the valence band of the HTM, the necessary exciton diffusion length to allow a quantum efficiency of 100 % would be 15 nm.

This size-reduction in two dimensions allows a higher fraction of excitons to reach the charge separating interface with TiO_2. Thus, this technique is suitable to further reduce the J-aggregate size. This coarse-porous structure filled with size-confined P2 J-aggregates is already applicable for solar cells if placed on a conductive substrate.

Also the application of such P2 J-aggregates on flat surfaces may be considered for solar cells, as they might absorb sufficient light within a densely packed monolayer of J-aggregates. A more densely packed J-aggregate layer could be achieved on the flat glass substrate next to a T300 layer, as shown in Fig. 3.37.

This higher density of P2 J-aggregates on the flat glass area next to the T300 layer is compared in Fig. 3.38 with the one coated similarly, but on a blank glass substrate, without any additional TiO_2 layer.

On the flat glass surface next to the coarse-porous TiO_2 layer (Fig. 3.38b) the J-aggregates were smaller and more densely packed than on the glass substrate, which had no additional TiO_2 layer in the center. For the size comparison it has to be taken into account, that the sample containing also the TiO_2 layer was sputtered with a 21-nm thicker gold layer in order to avoid charging during investigations of the TiO_2 layer. The difference in the aggregate distribution may originate from a

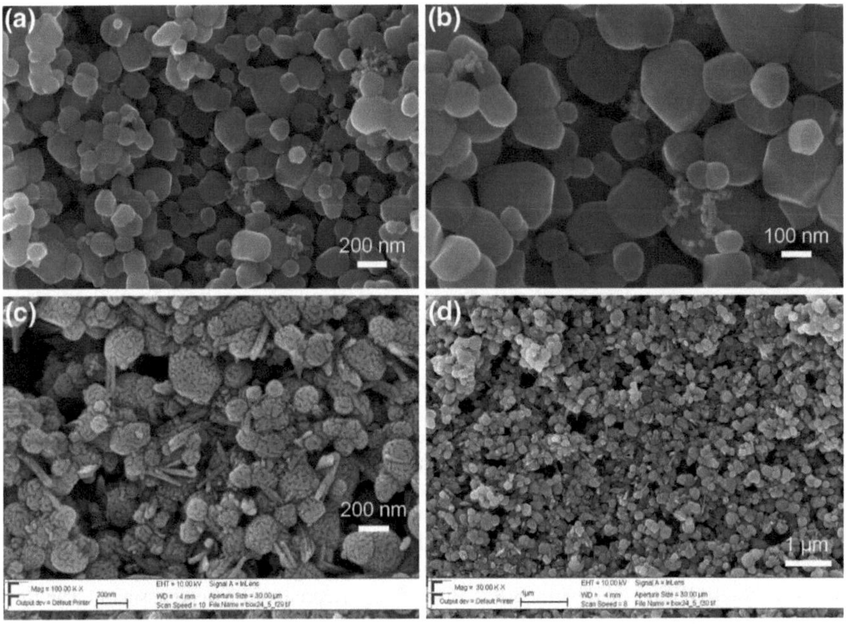

Fig. 3.36 SEM micrographs of reorganized P2 J-aggregates within a coarse TiO_2 layer. **a**, **b** Blank TiO_2 layer T300 after sintering at 450 °C for 1.5 h, in-lens. The nanoparticles in between should have reduced the sintering temperature. EHT: 10 kV, WD: 4 mm, aperture: 30 µm. **c**, **d** Similar T300 layer, but sintered at 500 °C for 1 h, infiltrated with rod-shaped J-aggregates of P2 after the reorganization in *n*-heptane-vapor and sputtering with 27-nm gold. The sample was prepared as described in the corresponding absorption spectrum in Fig. 3.35, but for these SEM investigations the sample needed to be sputtered with the gold to avoid charging on the glass substrate. **a**, **c** Marek PL, Hahn H, Balaban TS (2011) Energy Environ Sci 4:2366–2378 – Reproduced by permission of The Royal Society of Chemistry

different thickness of the initial amorphous P2 layer. Although both samples have been prepared with the same solution and coating parameters (6 mM in THF, 10 µl spin-coated at 3000 rpm) and even have been annealed together, hence under identical conditions in the same atmosphere. The P2-THF solution may have concentrated during the spin-coating. process, where the solution was dropped onto the TiO_2 layer of the fully accelerated sample. The fluid might have moved slower through the porous layer until it reached the flat surface, hence with a higher concentration, what might have resulted in a thicker amorphous P2 layer beside the TiO_2. An other reason for the different morphology might be, that the solution washed off some contaminants from the TiO_2 layer, water for example, and hence had a slightly different composition, which might affected the reorganization process.

However, it can be concluded, that with this dense coverage of small and narrow dispersed J-aggregates, even onto flat surfaces, the first step was successfully achieved on the way to make P2 applicable for solar cells. The coarse-porous TiO_2 layer allowed for a further decrease of the aggregate size and

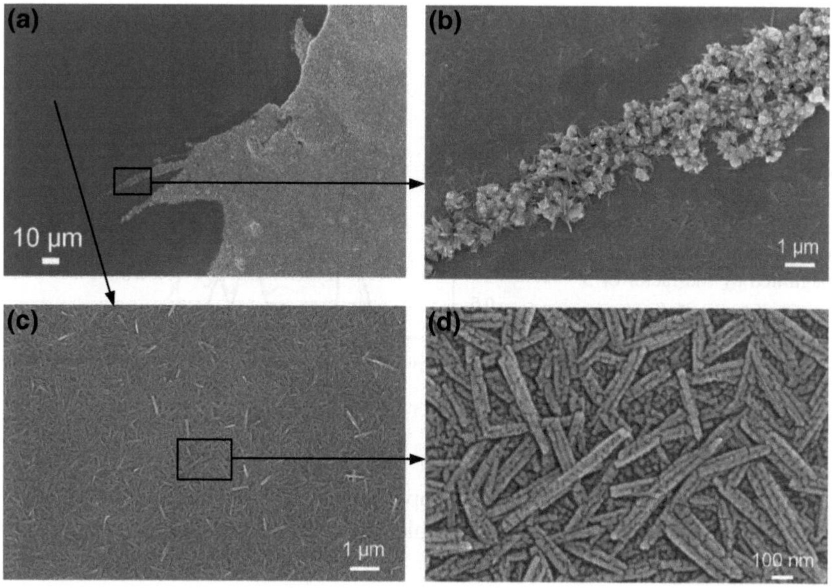

Fig. 3.37 SEM micrographs of reorganized P2 J-aggregates on glass, next to the TiO_2 layer. Onto the flat glass area next to the coarse-porous TiO_2 layer a higher density of the P2 J-aggregates can be seen. The surface roughness originates by the 27-nm thick gold sputtering

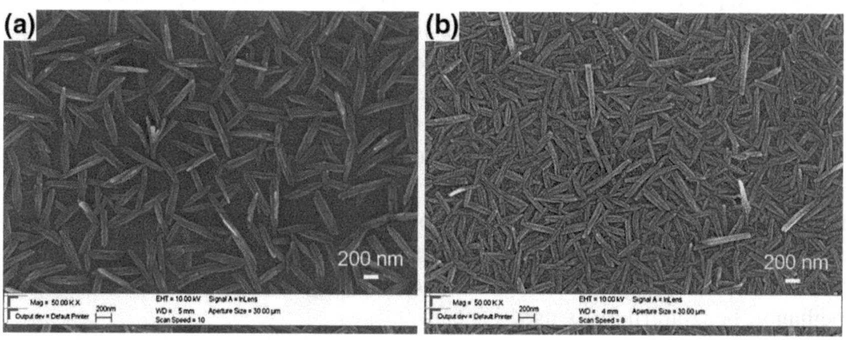

Fig. 3.38 SEM micrographs comparing different densities of reorganized P2 J-aggregates. **a** This sample from Fig. 3.31, shows the J-aggregate distribution on a bare glass substrate. Sputtered with 6 nm Au (**b**). The sample from Fig. 3.37, shows the J-aggregate distribution next to the TiO_2 layer after spin-coating the same amount of the 6 mM P2-THF (10 μl). It was sputtered with a 27-nm thick Au layer (**a**). Marek PL, Hahn H, Balaban TS (2011) Energy Environ Sci 4:2366–2378 – Reproduced by permission of The Royal Society of Chemistry

absorbance while maintaining the highly ordered J-aggregate structure and a short exciton diffusion path toward the TiO_2 interface.

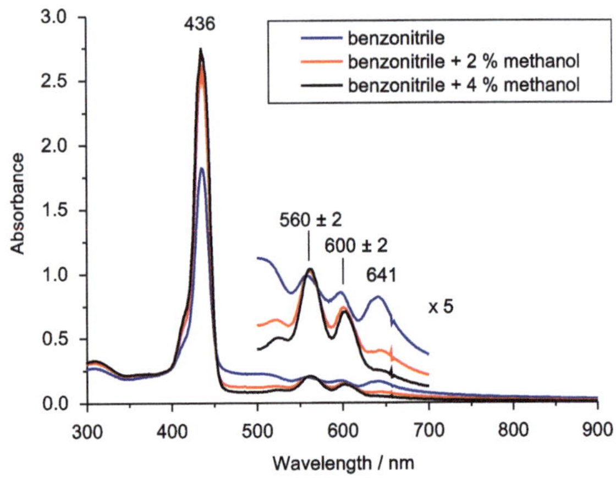

Fig. 3.39 Absorption spectra of P2 within benzonitrile. Absorption spectra of P2 within anhydrous benzonitrile including 0.05 M of (Bu_4NPF_6) (*blue* line) and after addition of 2 % (*red* line) and 4 % of methanol (*black* line). The inset shows the Q-bands' region magnified by the factor of 5

In the following the electrical properties of P2 and its J-aggregates were investigated in order to find a suitable material combination for the charge separation.

3.8 Cyclic Voltammetry

Cyclic voltammetry measurements have been performed in cooperation with Jean Paul Gisselbrecht in Strasbourg in order to determine the HOMO and LUMO level of P2. The studies could be performed within dichloromethane (DCM) and benzonitrile. As the supporting electrolyte, tetrabutylammoniumhexafluoro-phosphate (Bu_4NPF_6), was used. In order to distinguish the measured redox potentials of P2 monomers and those of P2 J-aggregates, the dispersion containing the J-aggregates has been measured first, followed by the addition of methanol, which disrupts the aggregates to monomers, and followed by a second measurement of this solution with the dominating fraction of monomers. As this measurement could not be performed in DCM due to an electrochemical degradation in combination with methanol, benzonitrile was used instead.

Figure 3.39 shows the absorption spectrum of P2 dissolved/dispersed in benzonitrile with and without methanol.

The absorption spectra of P2 in benzonitrile, including the supporting electrolyte $Bu_4N^+PF_6^-$, reveal the presence of J-aggregates in form of the red-shifted Soret band (shoulder) and the red-shifted Q-band at 641 nm, which could be decreased by the addition of methanol, This proves the disassembly of P2 J-aggregates to monomers.

The cyclic voltammetry measurements with and without the methanol however, could not confirm this change in the energy level of the Q-band (Fig. 3.40). In

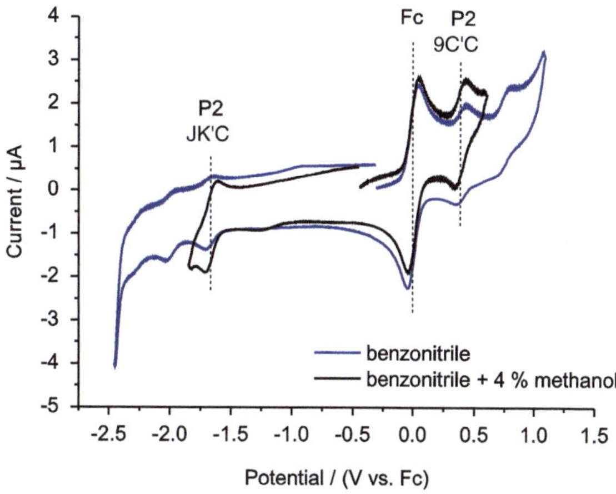

Fig. 3.40 Cyclic voltammetry of P2 within benzonitrile with and without methanol. Cyclic voltammetry of P2 in the presence of ferrocene (Fc) in anhydrous benzonitrile + 0.05 M $Bu_4N^+PF_6^-$ (*blue* line) and after addition of 4 % (v/v) methanol (*black* line)

benzonitrile with and without methanol the HOMO was measured to 0.38 V and 0.39 V and the LUMO to −1.66 V and −1.67 V, respectively.

While the HOMO-1 and the LUMO + 1 could be detected without the addition of methanol, after the addition of methanol only the first HOMO and LUMO could be measured again, probably due to reactions with methanol at higher potentials. It can be seen, that the HOMO and LUMO positions of P2 remained almost unchanged after the addition of methanol, but the peaks got sharper and more pronounced. This is an indication that these energy levels correspond to the P2 monomers, as their concentration increased. Similar values have been also obtained within anhydrous DCM solutions/dispersions with almost no change upon the addition of *n*-heptane, which was added to decrease the concentration of monomers and increase the J-aggregate concentration. Thus, it is assumed, that the measured redox potentials correspond only to the P2-monomers and not to the J-aggregates. The monomers could have dominated the redox-process, due to their higher diffusion coefficient within the solution.

The HOMO and LUMO levels of P2 monomers could be determined by cyclic voltammetry measurements within benzonitrile to: −5.2 and −3.1 eV vs. the vacuum level, respectively. The reference level of ferrocene (HOMO), which was used to calibrate the energy scale, is 4.80 eV vs. the vaccuum level [15].

As J-aggregates of P2 have a HOMO–LUMO gap, which is about 0.1 eV smaller than that of monomers and because only the LUMO splits due to the exciton coupling, the HOMO and LUMO levels of P2 J-aggregates can be estimated to: −5.2 and −3.2 eV, respectively.

The LUMO of P2 J-aggregates (−3.2 eV) is comparable with that of the standard ruthenium dye N719 [16] and fits well to the conduction band-edge of TiO_2, which is at about −4.2 eV [17]. The energy distance between these two levels of 1 eV offers a sufficient energy and driving force to effectively split the excitons at this interface, by an electron injection into the conduction band of

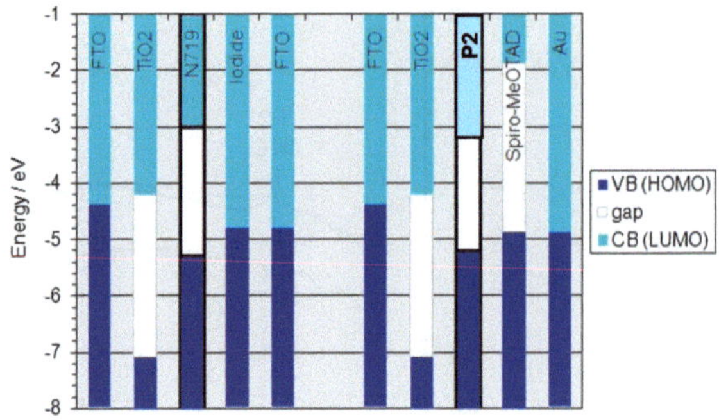

Fig. 3.41 Energy levels of the materials used for our solar cells. This graph shows the energy levels of the materials typically used for a DSSC with the standard ruthenium-dye N719 on the left side, where the Fermi-level of the second FTO electrode is pinned to the iodide/triiodide redox couple of the electrolyte (denoted by iodide). The right half of the graph shows the solid-state version (ss-DSSC) with the P2 J-aggregates and the solid hole transport material spiro-MeOTAD, which also pins the Fermi-level of its adjacent electrode, gold in this case. Marek PL, Hahn H, Balaban TS (2011) Energy Environ Sci 4:2366–2378 – Reproduced by permission of The Royal Society of Chemistry

TiO_2. The HOMO level does also match that one of the common hole transport material spiro-MeOTAD, which is about −4.8 eV [18], and is described in Chap. 6.6.3. Thus, this material combination was chosen for our solar cells together with the electrode materials FTO and gold, as shown in the right part of Fig. 3.41.

The energy level matching between the P2 aggregate's LUMO and the conduction band level of TiO_2 could be also proven by contactless photosensitization measurements of the P2-TiO_2 junction, which have been performed in cooperation with Huijser et al. [11]. For this measurement P2 J-aggregates were spin-coated from DCM onto a smooth TiO_2 layer on a quartz substrate. The sample was irradiated with a pulsed monochromatic laser in the visible wavelength range, while the conductivity of the TiO_2 layer was measured contactless by the microwave absorption technique, the so-called time-resolved microwave conductivity (TRMC). From the conductivity increase of the TiO_2 it can be inferred how many electrons got injected into the TiO_2 conduction band per incident photon (with energies smaller than the TiO_2 band gap), which is called the incident photon to charge separation efficiency (IPCSE). It reached 2 % at an excitation wavelength of 470 nm (monomeric absorption). As only one-fifth of the incident photons got absorbed this corresponded to an absorbed photon to charge separation efficiency (APCSE) of 10 %. The IPCSE at an excitation of the J-aggregates at 470 nm was around 0.4 %, corresponding to over 2 % APCSE. The reason, why this value was so far from unity, was the large aggregate size, with a mean diameter over 100 nm, thus exceeding the expected exciton diffusion length by far. Also the inappropriate aggregate distribution limited the electron injection rate, because many aggregates were piled over each other without a direct contact to the TiO_2. However, the electron injection from P2 aggregates to TiO_2 was much higher than to SnO_2 (undoped FTO), although the latter one has a higher electron affinity. Thus, TiO_2 was considered to be a suitable electron acceptor for sensitization with P2.

3.9 Dielectrophoresis

The dielectrophoresis (DEP) technique offers the possibility to manipulate different types of dispersed particles selectively in dependence to their permittivity and conductivity and align them between two electrodes. This method was applied in cooperation with Ralph Krupke at the KIT, in order to align the P2 aggregates across two electrodes for conductivity measurements. Beside the diffusion length of excitons, also the conductivity of P2 J-aggregates is of interest, because after the exciton dissociation and electron injection into the TiO_2 the remaining hole needs to move through the aggregate to reach the interface with the hole transport material. Additionally, dielectrophoresis could have the potential to separate the aggregates and, thus, allow to study the electrical and optoelectronic properties of, in the ideal case, monodisperse aggregates.

Since the pioneering work of Pohl H.A. [19, 20], dielectrophoresis has become a powerful technique to control the movement of diverse particles and fluid media by nonuniform electric fields. Depending on, whether the permittivity of a dispersed particle is higher or lower than that of the liquid medium, it will be attracted toward higher electric field concentrations or repulsed, respectively. As the permittivity and the conductivity depend on the frequency of the applied alternating current (AC), it allows for a better tuning of the applied forces then the electrophoretic deposition, which is done with direct current (DC). With AC-fields the permittivity ratio between the particles and the fluid medium can be varied and with it the interactive forces. This allows to even switch between attraction and repulsion of dispersed particles. The Clausius–Mossotti function (*K*) describes the corresponding permittivity ratio for different particle geometries, like for cylindrical particles as follows:

$$K(\omega) = \frac{\tilde{\epsilon}_p - \tilde{\epsilon}_m}{\tilde{\epsilon} + 2\tilde{\epsilon}_m} \text{ with } \epsilon = \tilde{\epsilon} - i\frac{\sigma}{\omega}$$

where $\tilde{\epsilon}$ is the complex permittivity with the indices p for particle and m for medium, σ the conductivity, ω the angular frequency, and i the complex number. The real part of *K* describes the in-phase component of the effective polarizability of the particles. For positive real parts of *K* attractive forces act on the suspended particles toward higher electric field concentrations, what is called the positive DEP, whereas negative real parts of *K* lead to a repulsion of those, the so called negative DEP [21, 22].

In addition to the DEP effect also electrothermal effects can occur, which may dominate over the DEP effect at certain conditions. These effects are the so-called electrothermal flow and the AC electroosmosis.

The AC electroosmosis causes a steady, one directional, motion of the fluid parallel to an asymmetric pair of microelectrodes due to the tangential component of the non-uniform AC-field, which interacts with ions in the diffusion layer. The electrothermal flow is caused by a thermal gradient, which originates by the Joule heating of the electrolyte solution, which leads to a temperature dependent gradient in the conductivity and permittivity of the solution. This effect causes vortices within the fluid, similar to thermal convection above a heating element, but in contrast to the thermal convection this effect occurs on a much smaller, a micrometer scale, and thus is of relevance for microsystems. The vortices enhance the accumulation of particles in front of the edge to one microelectrode. The review from Cao et al. presents an overview about such electrohydrodynamic effects, including also direct current (DC)-effects [23].

3.9.1 Dielectrophoretic Deposition on Interdigitated Electrodes

An interdigitated gold electrode structure has been chosen to deposit P2 aggregates across a gap of 1.6 μm width between gold electrodes, with had the width of 1.9 μm and a height of 110 nm. The electrode structure was prepared by standard photolithography on a 0.1 mm thick quartz glass [24]. To identify suitable parameters for attracting P2 aggregates from dispersions in dichloromethane the effects of dielectrophoresis have been observed in situ through an optical microscope. Two identical interdigitated electrodes (IDEs) were present on one chip, but only one was supplied by an alternating electrical field to induce an aggregate deposition and alignment by positive DEP. The field of (± 80 V) at 100 kHz was pulsed with 100 μs/ms in order to not heat up the solution significantly, as this could happen at such high potentials and lead to side-effects like boiling the solvent. This potential was necessary to see an effect through the optical microscope during the aggregate deposition from a drop (40 μl) of a 0.04 mM dispersion of P2 in DCM. The drop dried within a few tens of seconds on the chip during the applied potential. Figure 3.42 shows the results of the dielectrophoretic deposition onto the active area of the IDE.

This active IDE area can clearly be distinguished from the surrounding area, due to the DEP enhanced aggregate deposition in contrast to the control area on the left-hand side (a). The SEM micrographs show a predominant alignment of the aggregates perpendicular to the gold electrodes, like visible to the right of the borderline of the active IDE area in b. The success of the dielectrophoretic alignment was most significant in the area of the maximum field concentration at the ending tips of the electrode fingers (c,d).

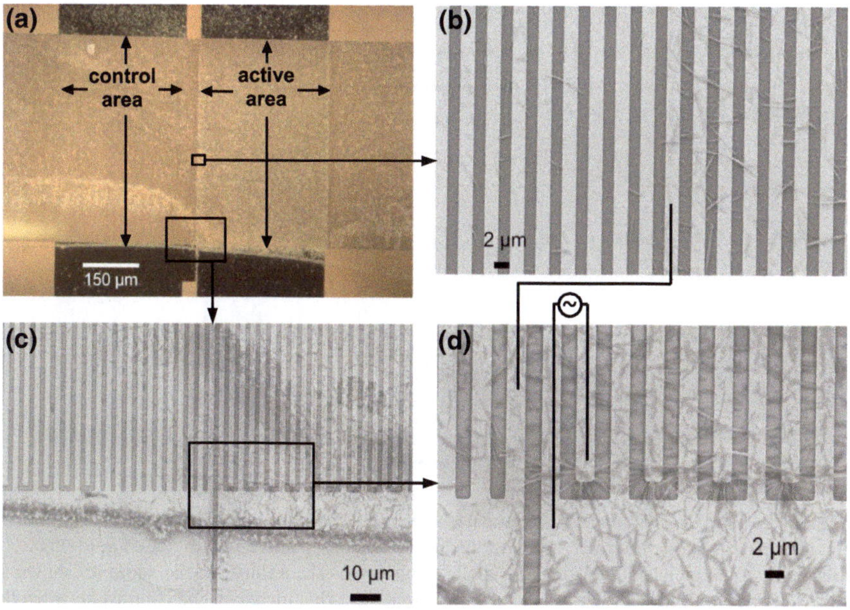

Fig. 3.42 P2 aggregates aligned by dielectrophoresis onto an interdigitated gold electrode. **a** The optical microscope image shows the chip with two identical interdigitated electrodes (IDEs) on the insulating quartz glass substrate, of which only the right hand was active, by an applied AC field during aggregate deposition. The one on the left served as a control. Note, that the borders of the active area can be seen due to an increased aggregate deposition, in contrast to the control area. The IDE areas reached to the tips of the arrows and were surrounded by similar gold structures but not interdigitated, like the 40-µm wide stripe between both IDEs. **b** SEM micrograph of the interface between the active and the not active area. These SE2 detector images give the aggregates a high morphology contrast and the substrate a high material contrast, where the quartz glass appears dark and the gold bright. Although the samples were not sputtered with any conductive coating for these images, the quartz area did not charge, due to the narrow gaps of 1.6 µm. The SEM micrographs in **c** and **d** show the bottom part of the IDE str, Bürck J, Garab G, Hahn ucture, and were recorded in a mixed-detector mode, where the SE2 and the in-lens detector were used to 50 % each, during the image recording. Thus, the aggregates appear darker than in the SE2 micrograph in **b**, what gives small aggregates a higher contrast without loosing too much of the material contrast of the substrate, which comes mostly from the SE2 detector. EHT: 5 kV, WD: 5 mm, aperture: 30 µm. Reprinted with permission from Chappaz-Gillot C, Marek PL, Blaive BJ, Canard GH, Jávorfi T, Kelemen L, Krupke R, Mössinger D, Ormos P, Reddy CM, Roussel C, Steinbach G, Szabó M, Ulrich AS, Vanthuyne N, Vijayaraghavan A, Zupcanova A, Balaban TS (2012) J Am Chem Soc 134:944-954. Copyright 2012 American Chemical Society

Beside the orientation across the electrodes it was also of interest, if it will be possible to separate certain aggregates, for example dependent to their length or diameter, and thus achieve a more monodisperse distribution. This would allow for a better correlation of the following electrical conductivity measurements to the size of the aggregates, in contrast to different thick or different long aggregates

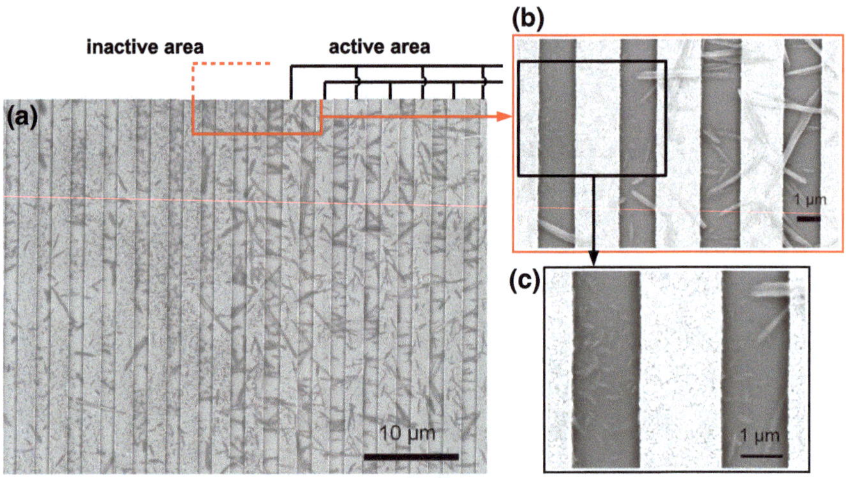

Fig. 3.43 SEM micrograph of P2 aggregates separated by size via dielectrophoresis. These SEM micrographs show the interface between the active and the inactive interdigitated gold-electrode area with P2 aggregates deposited by dielectrophoresis. The sample is the same as shown in Fig. 3.42, but imaged with the in-lens detector alone (**a**). This increased the contrast between the aggregates and the substrate, but decreased the material contrast between gold and quartz. Therefore, the drawing over the top right corner illustrates, where the gold stripes were and which of them were contacted for the dielectrophoretic deposition (active area). **b** and **c** show magnifications of the marked areas, which were recorded in a mixed-detector mode (in-lens:SE2 = 1:1) for a better contrast of the aggregates and the substrate (gold and quartz). EHT: 5 kV, WD: 5 mm, aperture: 30 μm. Reprinted with permission from Chappaz-Gillot C, Marek PL, Blaive BJ, Canard G, Bürck J, Garab G, Hahn H, Jávorfi T, Kelemen L, Krupke R, Mössinger D, Ormos P, Reddy CM, Roussel C, Steinbach G, Szabó M, Ulrich AS, Vanthuyne N, Vijayaraghavan A, Zupcanova A, Balaban TS (2012) J Am Chem Soc 134:944-954. Copyright 2012 American Chemical Society J, Garab

bridging the gap in different angles, what is the case after conventional deposition techniques without DEP.

It could be observed, that indeed a separation of different aggregate-sizes occurred as larger aggregates predominantly accumulated on the active area, due to the positive DEP effect, and very small aggregates remained on the inactive area. Figure 3.43 shows again the same interface between the active (right-hand side) and the inactive (left-hand side) area with optimized SEM parameters to see also very small aggregates.

This representative view in Fig. 3.43a, of the same IDE as seen in Fig. 3.42, gives a good overview over the size-distribution of the aggregates. It was recorded with the in-lens detector, which gives the aggregates a higher material contrast and thus makes also very small aggregates visible in a relatively low magnification. As the in-lens detector does not give a good material contrast between gold and quartz, the drawing in the upper part indicates which stripes were the gold electrodes and which were contacted, thus active during DEP. Whereas larger aggregates accumulated in the active electrode region with a high degree of orientation perpendicular to the active electrode fingers, the smallest aggregates were predominantly found on the inactive area (a-c, left sides). The first gold finger in the inactive area, that is, the first not contacted electrode in the drawing (a, top), can be regarded as part of the active area too, because a strong field gradient existed also between this electrode and the first biased one.

It can be seen, that the DEP technique is a powerful technique to manipulate such aggregates. It was indeed possible to align the aggregates predominantly perpendicular across the electrodes, allowing for more precise conductivity measurements due to a more monodisperse aggregate length bridging the gap. Additionally, this technique revealed a high capability to sort the aggregates macroscopically by size. By optimizing the DEP field parameters and the IDE gap sizes an even higher monodispersity of the deposited aggregates might be possible. This would allow to measure the optoelectronic properties of the aggregates with a better correlation to their morphology.

3.9.2 Controlling the Aggregate Size by Dielectrophoresis

By further adjustments of the dielectrophoretic field parameters it could be observed, that it was indeed possible to further increase the size-selectivity and achieve a deposition of exceptional small and almost monodisperse aggregates without the typical big aggregates, which normally dominated the morphology by conventional deposition techniques.

The new field parameters were: frequency: 300 kHz, potential: ± 80 V, pulsing: 100 μs/ms. The used P2 dispersion was 0.15 mM in anhydrous DCM, and the gap size of the IDE was: 1.6 μm.

In a cooperation with Jakub Pšenčík from the Charles University in Prague also the deposition of natural chlorosomes was tried by DEP. The used dielectrophoretic deposition parameters however, did not significantly affect the deposition of the natural chlorosomes. This can be also attributed to the difference in the solvents, which was water for the chlorosomes in contrast to DCM used for the P2 aggregates. Suitable field parameters to manipulate the deposition of the tiny chlorosomes, between the relatively large electrode gaps of the IDEs could not be found. Thus, the chlorosomes were deposited unaffected by the dielectrophoretic field only for comparison reasons and investigation of their size distribution. Figure 3.44 shows the significantly size-reduced P2 aggregates as deposited under the optimized DEP parameters in comparison with the natural chlorosomes.

The P2 aggregates, which could be deposited at 300 kHz (Fig. 3.44a, b) were significantly smaller, than those which could be achieved with any other deposition technique directly from DCM, except the low defined aggregate deposits by spraying. The narrow size-distributed and the low content of larger aggregates is remarkable. The spherical bright dots attached to the gold electrodes (a) are a residue of larger aggregates from a previous experiment, which have not been removed completely. The substrate has been re-used for this experiment, after checking that

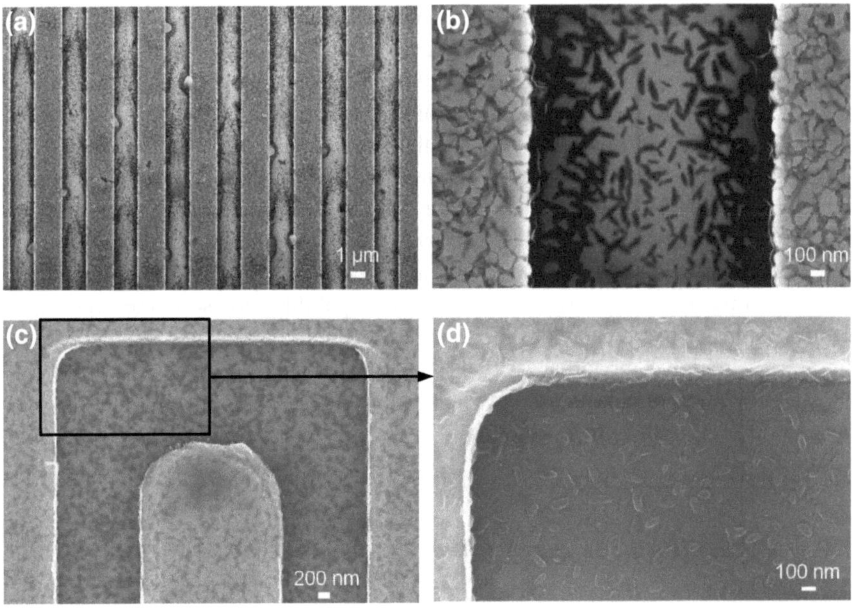

Fig. 3.44 Size-reduced and almost monodisperse P2 aggregates and natural chlorosomes on IDEs. **a**, **b** P2 aggregates have been deposited from a 0.15 mM dispersion in anhydrous DCM at 300 kHz, ± 80 V and 100 µs/ms pulsing onto an interdigitated electrode (IDE) with 1.6 µm gap size. The big globular particles attached to the gold electrodes did not occur during this deposition, but remained form a previous one. **c**, **d** Chlorosomes from the species C*hlorobaculum tepidum* on a similar IDE with 1.3 µm gap size. The dielectrophoretic field had no significant effect on the chlorosomes. In-lens, EHT: 10 kV, WD: 2 mm, aperture: 20 µm (**a**, **b**) and 30 µm (**c**, **d**)

only these spots remained from the previous experiment. Under the optical microscope it could be clearly seen, that the small aggregates accumulated on the active IDE area, thus showed a positive DEP effect. A deposition of comparable small and almost monodisperse distributed P2 aggregates without DEP was only possible by the reorganization method from amorphous P2 layers, a technique which has been developed in parallel. As it is known, that the change from positive to negative DEP, that is, from attraction to repulsion, occurs for smaller particles at higher frequencies then for larger particles [25] it is expected, that this crossover point was reached for the larger P2 aggregates at 300 kHz, but not for the small ones. Thus, larger P2 aggregates probably got repelled from the microelectrode structure, whereas the smallest aggregates got attracted. An other explanation or an additional effect may be also a perturbation of the agglomeration process during the deposition, what may have caused a higher concentration of small aggregates.

By this technique P2 aggregates could be deposited, which are of a comparable size as the natural chlorosomes, shown in (c, d), which are expected to have an exciton diffusion length in the same order of magnitude than P2. With these field parameters it could be shown, that the DEP technique has a high potential to control the deposition of such self-assemblies. This technique might be also applicable in future for the deposition of monodisperse and size reduced P2 aggregates on substrates for solar cells, like on interdigitated back contacted solar cell architectures. However, it is not possible to upscale economically interdigitated electrode structures with such narrow electrode gaps, yet. Therefore, the direct deposition of P2 J-aggregates will not be applied to produce solar cells. The method of choice for this purpose will be the two step process, which has been developed in parallel and allows to form small J-aggregates from amorphous layers economically and scalable.

Using the DEP technique a focus was put on the deposition of larger aggregates, which could bridge the electrode gap for the following conductivity measurements. In order to find a better correlation between the aggregate morphology within the gap and the electrical properties, the chip structure was changed to one with a single gap, which could be bridged only by a small number of aggregates. A chip design, which is often used for a similar purpose, the deposition of single carbon nanotubes, involves so-called floating electrodes.

3.9.3 Dielectrophoretic Deposition on Floating Electrodes

For a precise conductivity measurement it is important to know the number of aggregates bridging the electrodes, as well as their length and diameter. As this was difficult to estimate in the large IDE structure, another electrode design was used in order to deposit and characterize only a few or a single aggregate bridging the electrodes. A common technique to realize this for the characterization of single carbon nanotubes is the dielectrophoretic deposition between so-called floating electrodes. There a metal electrode structure is applied onto a conductive substrate with a thin insulating layer in-between. The AC-field is applied between the metal structure and the conductive substrate, which is commonly a highly doped silicon wafer with an insulating SiO_2 layer on top. The floating electrodes are subdivisions of the contacted metal structure and isolated from it by narrow gaps. As they are not contacted, they couple inductively mainly with the substrate, but also over the gap with the contacted metal electrode. Once a carbon nanotube bridges the gap, attracted by DEP, the floating electrode gets electrically contacted with the biased metal electrode. This eliminates the field gradient and no further carbon nanotubes get attracted into the gap anymore. Afterward those floating electrodes which got bridged by a carbon nanotube of interest are being bonded, that is, contacted too, in order to apply a current through the nanotube bridging the gap.

For the experiments to deposit P2 aggregates across the gap to the floating electrodes a microelectrode structure of palladium has been applied by electron beam lithography onto the SiO_2 layer on a highly doped silicon substrate. The main Pd electrode structure was contacted to the AC supply and branched off to both sides, where the end sections were separated from that electrode by different sized gaps, as shown in Fig. 3.45a. These end sections were not contacted and not

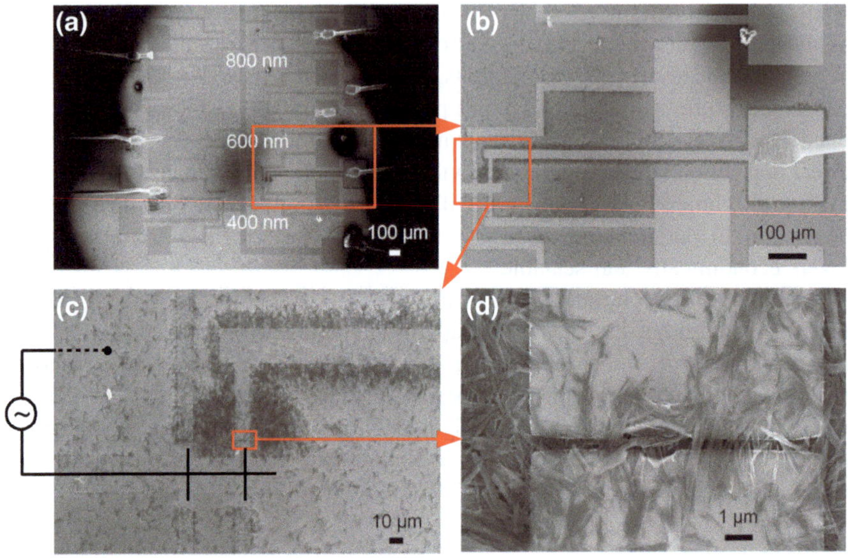

Fig. 3.45 SEM of P2 aggregates deposited by dielectrophoresis around floating electrodes. The partly contacted palladium structure is applied on an insulating 800-nm thick SiO_2 layer on a doped silicon wafer. The main Pd electrode (**a**), in the middle, ranging from the top to the bottom right corner) branches off to both sides with four floating electrodes of Pd at each branch (**b**, **c**), which are separated from the main Pd electrode by small gaps: the upper eight floating electrodes are separated by 800 nm gaps, the middle ones by 600 nm and those at the bottom by 400 nm gaps (right branch: **b–d**). For dielectrophoresis only the main Pd electrode and the silicon substrate has been contacted to the alternating current (AC) supply like drawn in **c**. The gap to the floating electrode, surrounded by the highest aggregate accumulation, is magnified in d. In-lens, EHT: 10 kV, WD: 5 mm, aperture: 30 µm. Reprinted with permission from Chappaz-Gillot C, Marek PL, Blaive BJ, Canard G, Bürck J, Garab G, Hahn H, Jávorfi T, Kelemen L, Krupke R, Mössinger D, Ormos P, Reddy CM, Roussel C, Steinbach G, Szabó M, Ulrich AS, Vanthuyne N, Vijayaraghavan A, Zupcanova A, Balaban TS (2012) J Am Chem Soc 134:944-954. Copyright 2012 American Chemical Society

even bonded during the aggregate deposition. They formed floating electrodes in which the electric field was mainly induced from the contacted silicon substrate beneath the 800-nm thick SiO_2 insulator. All in all 24 floating electrodes were present on the chip of which each eight had the same gap size to the main Pd electrode, namely 800, 600, and 400 nm.

The visible wires at some floating electrodes (a, b) have been bonded to the pads after the aggregate deposition, only for conductivity measurements through the gap. The applied field for the DEP was $\pm$ 1 V at 300 kHz (not pulsed). 50 µl of a 3.6 mM P2 solution in dichloromethane have been poured into 500 µl of anhydrous *n*-heptane to induce precipitation. 40 µl of this solution have been dropped quickly onto the chip and after 4 min of dielectrophoresis the remaining solvent has been blown away with nitrogen, followed by a purge with clean *n*-heptane.

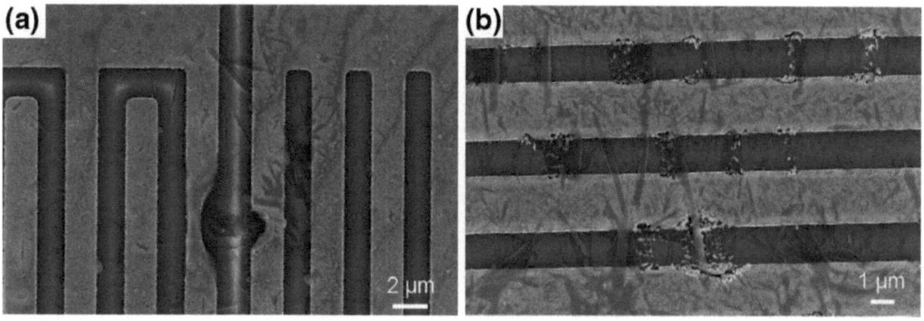

Fig. 3.46 SEM micrographs of metallic microelectrodes melted around P2 aggregates. Degradation of some interdigitated microelectrodes observed after dielectrophoretic deposition of P2 aggregates at ± 80 V within a 1.6 μm gap (**a**) and a 1.4 μm gap (**b**)

Whereas no significant P2 aggregate accumulation could be observed within the 800 and 600 nm gaps and around the corresponding floating electrodes, the floating electrodes with the 400 nm gaps were surrounded by a small layer of aggregates. Although the four floating electrodes, branching off in c were intended to have similar 400 nm gaps, one of them attracted much more aggregates, than the other three. The origin for this effect seems to correspond to the melted spot in the center of the magnified gap in d, as this could not be observed within the other 7 gaps with 400 nm size. Probably this gap was initially in this part even smaller or bridged by a higher conductive bundle of P2 aggregates. Similar melted spots could be observed also on some interdigitated electrode structures as shown in Fig. 3.46.

The spot at the interdigitated electrode (IDE) in Fig. 3.46a had a similar appearance as that within the 400 nm gap of the floating electrode in Fig. 3.45a. On an IDE with a smaller gap size (1.4 μm) many aggregates seem to have individually short circuited the electrodes along their long axis or facilitated sparks across the gaps.

A plausible explanation for the effect that the floating electrodes attracted the aggregates not only within the gap to the contacted electrode, but all around their borderline, is given in the following paragraph.

These effects, like the significant accumulation of the P2 aggregates all around the floating electrodes with 400 nm gaps (Fig. 3.45a), and especially around that one with the melted spot within the gap (b, d), may be related to electrothermal effects, rather than to pure dielectrophoresis. The alternating current electrothermal effect (ACET) is caused by a nonuniform heating of the liquid medium due to the non-uniform AC-field. These temperature gradients cause gradients in the permittivity and conductivity of the fluid what again leads to directed forces due to dielectrophoretic interactions. This results in vortices of the fluid in a certain direction depending on the electrode geometry, frequency and conductivity [26, 27]. These vortices wash up particles toward high electric fields and thus enhance their DEP induced deposition. Electrothermal effects are more pronounced in the presence of a static electric field in combination with an AC-field [28], what is supposed to be also

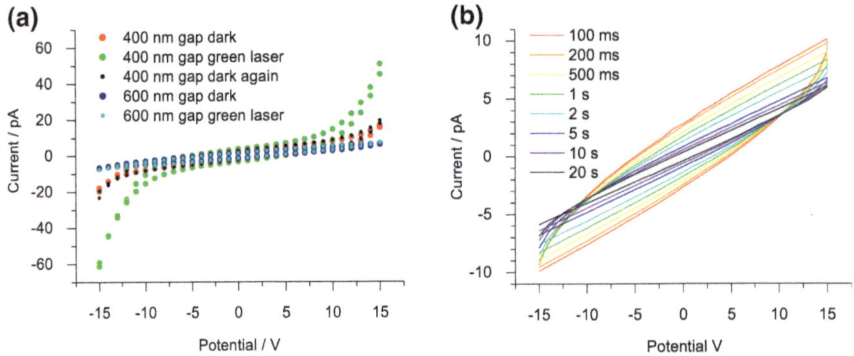

Fig. 3.47 Photoconductivity of P2 bridging the 400 and 600 nm Pd gaps. **a** The photoconductivities across the Pd gap to the floating electrodes from Fig. 3.45 were measured in the dark and during the irradiation by a *green* laser (1 mW, 532 nm, 3 mm^2). The integration time was 1 s per point, which has been determined as an optimum from the series in (**b**) recorded at the 600 nm gap in the dark. The integration time of 1 s was a compromise between the time to measure one I–V curve and the hysteresis. **a** The aggregates within the 400 nm gap, with the high accumulation of P2 aggregates (Fig. 3.45d), showed a significant increase in conductivity upon irradiation in contrast to the 600 nm gaps. The 800 nm gaps did not show any conductivity increase upon irradiation, too (not shown)

the case in our system. As the P2 aggregates within the metal gap show a diode characteristic (Fig. 3.47), this junction is supposed to have rectified the AC supply through the gap and led to a charging of the floating electrode. The assumed charging of the floating electrode may be explained with its always lower potential compared to the main Pd electrode, what may have led to a Schottky barrier, which is also asymmetric, thus causing a charging of either plus or minus sign. An indication for the dominance of electrothermal effects over dielectrophoresis, in our system, gives the gap with the melted spot, where obviously the thermal influence was maximal and so the aggregate deposition, extending to a radius of over 30 μm (Fig. 3.45c) around the gap. As the charging of the floating electrode should be dependent on the conductivity of the bridged gap, this effect is supposed to be the origin for the dependence on the gap-size of the observed aggregate accumulation around the floating electrodes, being most pronounced at the 400 nm gaps. This self-induced DC-biasing of the floating electrodes with small gaps seems to be the reason, why an increased aggregate deposition occurred only around the floating electrodes and not around the AC-supplied main Pd electrode, which was on the highest AC-field, but with no DC-offset.

It can be concluded that such floating electrode structures in combination with (organic) semiconducting particles may self-catalyze significantly electrokinetic effects at very low electric fields, like the ± 1 V (= 2 Vpp [peak to peak]) used here. On the interdigitated electrode structure a potential of ± 80 V was needed to achieve a reasonable P2 aggregate deposition. With an optimization of the electrode structure for these electrothermal effects using suitable floating electrodes or implementing a third electrode underneath the IDE structure, the aggregate deposition could be significantly improved and the necessary potentials reduced.

P2 aggregates have been also deposited onto the gaps of floating electrodes from more diluted DCM dispersions what allowed to bridge some gaps also by a single aggregate.

3.9.4 *Photoconductivity*

The conductivity of the P2 aggregates bridging the Pd gaps in Fig. 3.45 have been measured after bonding the floating electrodes at the pads and using as counter-electrode the main Pd electrode. The I–V-curves through the bridged gaps have been measured in the dark and during the irradiation with a laser of 1 mW at 632-nm wavelength and approx. 3 mm^2 spot size. Figure 3.47 shows that the conductivity increased upon irradiation of the 400 nm gap, but almost no changes were observed at the 600 nm gap.

The aggregates showed a non-linear increase in conductivity above a threshold of about 10 V. A strong hysteresis was observed for the I–V measurements, which has been investigated in more detail at the 600 nm gap in the dark (b). This needed to be compensated by a longer integration time for each measurement point and 1 s/point has been chosen for the measurements (a) as compromise between measurement time and accuracy. Whereas the conductivity of the 400 nm gap increased significantly upon irradiation, that of the 600 nm gap was hardly measurable and that of the 800 nm gap, bridged by almost statistically distributed J-aggregates from Fig. 3.45 not measurable at all.

The 400–800 nm gaps, which have been bridged by single aggregates from the diluted dispersion in DCM, showed also no measurable conductivity increase upon irradiation with the green laser.

Figure 3.48 compares the photoconductivity of the 400 nm gap, bridged by the high number of aggregates, with that of the interdigitated gold electrode from Fig. 3.42.

3.9.5 *Estimation of the Conductivity*

In order to estimate the conductivity of the aggregates the following parameters were used for the interdigitated gold electrode:

Width of the gap in the IDE: 50 mm (520 μm·335 μm/3.5 μm), gap length: 1.6 μm, estimate of the number of J-aggregates bridging the gap: 10 000, dimensions of the estimated average J-aggregate within the gap: 1.6 μm long and 0.1 μm·0.1 μm thick.

At higher potentials (12–15 V) and under irradiation with the green laser the conductivity of the P2 J-aggregates in the direction of their long axis is estimated

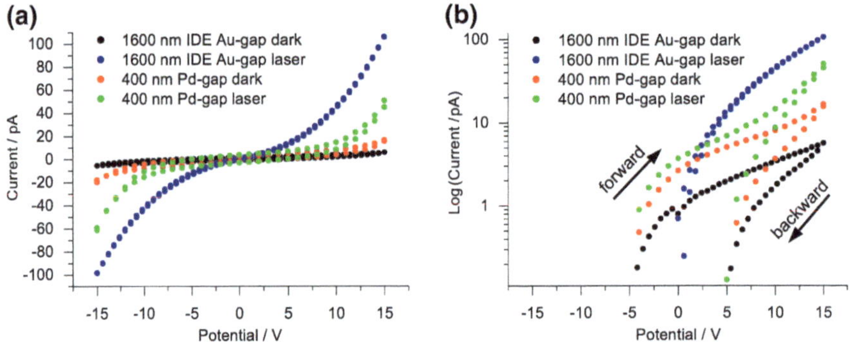

Fig. 3.48 Photoconductivity of P2 aggregates on the interdigitated gold electrode. These measurements compare the photoconductivity of P2 aggregates within the 1.6 μm wide gaps of the interdigitated gold-electrode (IDE) from Fig. 3.42, measured at 3 s/point (*blue* and *black* curves) with the P2 aggregates bridging the 400 nm gap of the palladium structure (*green* and *red* curves), shown in Fig. 3.45d, measured with an integration time of 1 s/point. The irradiation source was in both cases a *green* laser (1 mW, 532 nm, 3 mm^2 spot size). Whereas the I–V sweep of the 400 nm curves was started and finished at −15 V, that of the 1600 nm was started and finished at 0 V. The diagram on the right-hand side shows the same data in a logarithmic plot (positive current only). Left: Reprinted with permission from Chappaz-Gillot C, Marek PL, Blaive BJ, Canard G, Bürck J,Garab G, Hahn H, Jávorfi T, Kelemen L, Krupke R, Mössinger D, Ormos P, Reddy CM,Roussel C, Steinbach G, Szabó M, Ulrich AS, Vanthuyne N, Vijayaraghavan A, Zupcanova A,Balaban TS (2012) J Am Chem Soc 134:944-954. Copyright 2012 American Chemical Society

to be in the range of $2{\cdot}10^{-7}$ S/m. This corresponds to a resistivity of $5{\cdot}10^{6}$ Ω·m, which is about 3 orders of magnitude higher than that of undoped silicon. In the potential range up to 5 V the conductivity under this irradiation was about 5 times lower ($4{\cdot}10^{-8}$ S/m), corresponding to a resistivity of about $2.5{\cdot}10^{-7}$ Ω·m. Without irradiation the conductivity was within the potential range up to about 12 V approx. one order of magnitude lower ($4{\cdot}10^{-9}$ S/m).

For the aggregates bridging the 400 nm gap between the palladium gap on the floating electrode structure (Pd-FE) the conductivity under irradiation with the green laser was, within the potential range of 12–15 V, about 25 times higher ($5.6{\cdot}10^{-6}$ S/m), than that of the Au-IDE. Within the potential range up to approx. 5 V it was about 18 times lower ($3{\cdot}10^{-7}$ S/m, $3{\cdot}10^{6}$ Ω·m) than above 12 V and in the dark up to 12 V it could be approximated to $2{\cdot}10^{-7}$ S/m ($5{\cdot}10^{6}$ Ω·m).

It was estimated that the 400 nm long gap was bridged over the width of 10 μm by a densely packed array of J-aggregates with 0.1 μm · 0.1 μm cross-section, each and 0.4 μm length, thus by about 100 aggregates.

In both types of gaps the conductivity increases significantly at higher potentials. This can be explained by a space charge limited current (SCLC), which starts rising faster at a certain potential, when most trap states have been filled with charges, smoothing out the potential landscape within the aggregates. A significant difference between both types of gaps, or J-aggregate lengths, arises upon irradiation with the green laser. Whereas the conductivity of the 400 nm long J-aggregate bridge increases by a factor of less then two upon irradiation at low potentials, that of the 1600 nm long J-aggregate bridge increases by about one order of magnitude upon irradiation at low potentials. This can be explained also by a higher number of trap states within the longer J-aggregates which shield the electric field and thus dominate the resistance within a higher potential range until the irradiation helps to release the trapped charges leading to an increases of the conductivity. The logarithmic plot of the conductivity in Fig. 3.48 is better suited to see the threshold potential at which the traps are filled and hence the conductivity rises faster. This diagram confirms, that the threshold potential of the SCLC is reached at about 10 V for the 400 nm short aggregates (green and red curves) and is not reached up to 15 V for the 1600 nm long aggregates (blue and black curves).

In the logarithmic plot the hysteresis between the forward and backward potential sweep is also pronounced. It results from a relatively high capacitance in combination with a relatively low conductivity. Although the difference in the work function of the electrode materials Au and Pd is very small, it can not be excluded, that it contributed to the conductivity measurements by different charge injection rates. As the gold or Pd have their Fermi-level close to the HOMO of the aggregates it can be assumed, that this photoconductivity corresponds to a hole transport rather than to an electron transport, which may be hindered by a higher Schottky barrier. This agrees with additional measurements, where no occurrence of electroluminescence on the gold IDE chip could be detected, which was tried with a sensitive CCD camera through an optical microscope. This could also mean that an eventual electron transport occurs over triplet states and that the intersystem crossing into the singlet state is not likely. This would be beneficial for solar cells as it would reduce the recombination rate.

The conductivity of the aggregates might be not sufficiently high for an efficient charge transport through the entire long axis of the aggregates. However, the conductivity rose nonlinear with the decrease of the path length and hence might be significantly higher on a short length scale, which could not be investigated further with our electrode structures. However, P2 aggregates do not have such high requirements for the conductivity like dyes in organic or hybrid solar cells, because in our solar cell concept a third material takes over the part for the long-range hole transport. Using an additional hole transport material the maximum length for the hole transport within P2 aggregates would be in the range of the aggregate diameter and therefor its conductivity seems to be sufficiently high.

To study whether the photoconductivity correlates with the absorption spectrum when the irradiation wavelength is varied, a UV–Vis-NIR absorption spectrometer was used as monochromatic light source for conductivity measurements during the irradiation by a continuous wavelength sweep in the spectrometer. Because the irradiation intensity was very low, the integration time for each current-measurement point was set to 10 s and the duration for the whole wavelength sweep was set to 1.5 h. The spectral photoconductivity of the IDE sample with the bridged 1.6 μm gap is shown in comparison with a typical P2 J-aggregate spectrum in Fig. 3.49.

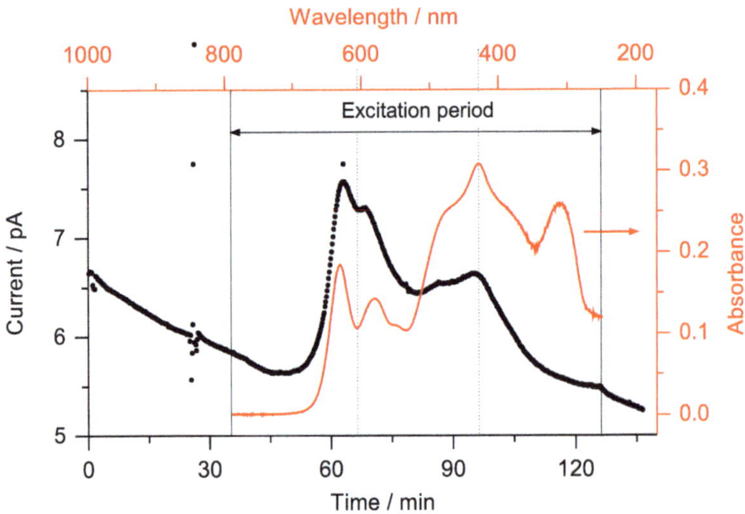

Fig. 3.49 Photoconductivity of P2 on the IDE within the UV–Vis-NIR spectrometer. The same interdigitated gold electrode bridged by the P2 aggregates, which has already be shown in Fig. 3.42 and on which the photoconductivity measurement from Fig. 3.48 has been performed, was used here again to measure the spectral photoconductivity using the monochromatic excitation source of the UV–Vis-NIR absorption spectrometer. A typical absorption spectrum of P2 J-aggregates is shown for comparison (*red*)

The spectral photoconductivity (dark dots) showed indeed a correlation with the absorption spectrum of P2 J-aggregates (red trace). At the beginning of the conductivity measurement in the dark a steady decrease can be seen. This trend of a decreasing baseline signal probably did not change when the wavelength sweep was started at 790 nm. When the excitation wavelength reached the first absorption band of the J-aggregates the conductivity steeply increased and decreased again after this Q band to reach a minimum between the two Q bands. Note the mirrored absorption spectrum due to the plotting in the direction of the time scale and not as usual in the direction of increasing wavelengths. The dotted vertical lines at wavelengths of 606 and 428 nm highlight the correlation at the point between the two Q bands and at the Soret band maximum. The steep decrease of the photocurrent at the beginning of the measurement in the dark shows the very slow relaxation time of the aggregate conductivity after an previous excitation. This slope in the baseline is the reason, why the spectral photoconductivity does not correlate better with the absorption spectrum. A similar slow relaxation time has been also observed by Schwab et al., who performed photoconductivity measurements on self-assemblies of meso-tetrakis(4-sulfonatophenyl) porphyrins, which bridged a 350 nm gap of a similar IDE [29]. However, the question remains, what causes this slow relaxation. It might be a high capacity.

The band at around 300 nm is the only one, which caused no change in the conductivity spectrum. This agrees with our hypothesis, that this band does not belong to the porphyrin's core absorption, but to the absorption of its side groups, the phenyl rings. As these are tilted by about 70° relative to the porphyrin plane, they do not belong to the extended pi-conjugated system of the porphyrin.

References

1. Balaban TS, Bhise AD, Fischer M, Linke-Schaetzel M, Roussel C, Vanthuyne N (2003) Angew Chem Int Ed Engl 42:2139–2144
2. Bhise AD (2005) A biomimetic approach for synthesizing artificial light-harvesting systems using self-assembly. PhD thesis, University of Karlsruhe. Forschungszentrum Karlsruhe GmbH, Karlsruhe, http://bibliothek.fzk.de/zb/berichte/FZKA7174.pdf
3. Takahashi K, Terashima T, Komura T, Imanaga H (1989) Bull Chem Soc Jpn 62:3069–3074
4. Balaban TS, Linke-Schaetzel M, Bhise AD, Vanthuyne N, Roussel C (2004) Eur J Org Chem 3919–3930
5. Balaban TS, Leitich J, Holzwarth AR, Schaffner K (2000) J Phys Chem B 104:1362–1372
6. Nerdel F, Schrader B (1961) Organische chemie. Walter De Gryter & Co, Berlin
7. Psencík J, Ikonen TP, Laurinmäki P, Merckel MC, Butcher SJ, Serimaa RE, Tuma R (2004) Biophys J 87:1165–1172
8. Balaban TS (2005) Acc Chem Res 38:612–623
9. Balaban TS, Linke-Schaetzel M, Bhise AD, Vanthuyne N, Roussel C, Anson CE, Buth G, Eichhöfer A, Foster K, Garab G, Gliemann H, Goddard R, Javorfi T, Powell AK, Rösner H, Schimmel T (2005) Chem Eur J 11:2267–2275
10. Balaban TS, Bhise AD, Bringmann G, Bürck J, Chappaz-Gillot C, Eichhöfer A, Fenske D, Götz DCG, Knauer M, Mizoguchi T, Mössinger D, Rösner H, Roussel C, Schraut M, Tamiaki H, Vanthuyne N (2009) J Am Chem Soc 131:14480–14492
11. Huijser A, Marek PL, Savenije TJ, Siebbeles LDA, Scherer T, Hauschild R, Szmytkowski J, Kalt H, Hahn H, Balaban TS (2007) J Phys Chem C 111:11726–11733
12. Mhuircheartaigh ÉMN, Blau WJ, Prato M, Giordani S (2007) Carbon 45:2665–2671
13. Marek P (2005) Selbstanordnende Zn-Porphyrine in makroporösen TiO_2-Schichten für farbstoffsensibilisierte Solarzellen. Diploma thesis, Technical University of Darmstadt
14. Marek PL, Sieger H, Scherer T, Hahn H, Balaban TS (2009) J Nanosci Nanotechnol 9:3708–3713
15. Pomrnerehne J, Vestweber H, Gun W, Muhrt RE, Bassler H, Porsch M, Daub J (1995) Adv Mater 7:551–554
16. Nazeeruddin MK, De Angelis F, Fantacci S, Selloni A, Viscardi G, Liska P, Ito S, Takeru B, Grätzel M (2005) J Am Chem Soc 127:16835–16847
17. Xu Y, Schoonen MAA (2000) Am Mineral 85:543–556
18. Kroeze JE, Hirata N, Schmidt-Mende L, Orizu C, Ogier SD, Carr K, Grätzel M, Durrant J (2006) Adv Funct Mater 16:1832–1838
19. Pohl HA (1958) J Appl Phys 29:1182–1188
20. Pohl HA (1978) Dielectrophoresis. Cambridge University Press, Cambridge
21. Ramos A, Morgan H, Green NG, Castellanos A (1998) J Phys D: Appl Phys 31:2338–2353
22. Zhang C, Khoshmanesh K, Mitchell A, Kalantar-Zadeh K (2010) Anal Bioanal Chem 396:401–420
23. Cao J, Cheng P, Hong FJ (2009) Sci China Ser E: Technol Sci 52:3477–3490
24. Krupke R, Linden S, Rapp M, Hennrich F (2006) Adv Mater 18:1468–1470
25. Wei M-T, Junio J, Ou-Yang HD (2009) Biomicrofluidics 3:012003
26. Lin Y, Shiomi J, Maruyama S, Amberg G (2007) Phys Rev B 76:045419_1-5

27. Chen DF, Du H (2006) J Micromech Microeng 16:2411–2419
28. Ng WY, Goh S, Lam YC, Yang C, Rodríguez I (2009) Lab Chip 9:802–809
29. Schwab AD, Smith DE, Bond-Watts B, Johnston DE, Hone J, Johnson AT, de Paula JC, Smith WF (2004) Nano Lett 4:1261–1265

Chapter 4
Conclusion: Biomimetic Light-Harvesting

It could be shown that the biomimetic self-assembling P2 aggregates have suitable properties for applications in solar cells. They self-assemble to J-aggregates, which have a highly ordered crystalline structure, what could be inferred from XRD and TEM investigations, their exciton coupling in the UV–Vis-NIR absorption spectra and also their ability to fluoresce in the solid-state. A high mean exciton lifetime in the J-aggregates of 40 ps could be measured by time-resolved fluorescence spectroscopy in cooperation with the group of Prof. Heinz Kalt at the KIT. With the dielectrophoresis technique new possibilities for the manipulation and alignment of P2 aggregates were demonstrated which can be used for characterizations like photoconductivity measurements, for sorting or for the deposition of such structures on solar cell substrates for example. The photoconductivity measurements showed promising optoelectronic properties, which might be useful also for other applications beside solar cells, like sensors, or for energy or data storage for example, as the aggregates seem to have a high capacity. The HOMO and LUMO levels could be successfully determined, proving a suitable fit to the conduction band edge of TiO_2 and to the common hole transport material (HTM) spiro-MeOTAD, which will be described in the next section.

Beside the determination of the optoelectronic properties of P2 the most important requirement for its applicability to solar cells was the limitation of the J-aggregate size, which should not exceed the expected mean exciton diffusion length at least in one dimension. Various methods could be applied to limit the aggregate size in a single deposition step, like the precipitation onto zinc oxide nanorod substrates, spraying of J-aggregate dispersions, drop-casting at low temperatures or a size-selective aggregate deposition using the dielectrophoresis technique. However, the results of all these methods were not satisfying for different reasons, but with a two step process we could obtain suitable results. Thereby a homogeneous amorphous layer of P2 has been first deposited from THF, a solvent, which hinders the self-assembly by coordinating the central zinc atom. In a second step, this layer was reorganized by inducing the self-assembly in a thermal post-treatment in *n*-heptane or its vapor phase. This solvent facilitates the self-assembly process without dissolving the dye completely. By optimizing this

P. L. Marek, *Biomimetic Dye Aggregate Solar Cells*,
Springer Theses, DOI: 10.1007/978-3-319-00636-9_4,

method, the deposited P2 J-aggregates could be significantly reduced in size. Additionally, a narrow distribution of different aggregate sizes and morphologies could be achieved as well as a homogeneous lateral distribution and alignment on the substrate. Whereas this technique worked fine on flat substrates, the reorganization to J-aggregates could not be achieved within mesoporous TiO_2-layers. When these layers were equipped with additional macropores, J-aggregates grew only within these voids at high concentrations of the infiltrated dye. Absorption spectroscopy measurements revealed that the major fraction of P2 thereby remained amorphous, most probably sitting invisible for SEM investigations within the mesopores. We could achieve an almost complete reorganization within coarse-porous TiO_2 layers, instead. Such layers helped to further reduce the aggregate size and offered the possibility to adjust the dye amount to higher light intensities without affecting the diffusion path length through the dye aggregates. The P2 infiltrated coarse-porous TiO_2 layers showed a broad absorption over almost the entire visible wavelength range from the UV region, limited by the absorbance of TiO_2, up to 645 nm. Such an architecture may be well suited for solar cells as it offers additional advantages, like an increased light scattering allowing for a higher absorbance, a lower surface area at which recombination can take place and an easier infiltration for various solid HTM.

Part II
Application of BLH to Solar Cells

Chapter 5
Theory and Literature Survey: Application of BLH to Solar Cells

5.1 Definition of Anode and Cathode in Solar Cells

The terms anode and cathode have not been used in this thesis, because it is not consistent for different types of solar cells and hence, would lead to confusion. Whereas in organic solar cells, for example, the negatively charged electrode is called anode, in dye sensitized solar cells (DSSCs) it is called cathode. The cathode (from Greek [káthodos], meaning: way down) is defined as that electrode, at which the reduction takes place. Hence, it is meant as source of electrons, which move down the potential gradient towards the anode. For the discharge of batteries and the most solar cells, this definition is true for the internal current, but not for the external one. For (solid-state) dye-sensitized solar cells, however, it is the other way around, maybe because there the external current flow direction has been taken for naming the electrodes.

5.2 Conventional Solar Cells

Conventional solar cells of the first generation are based on mono- or multicrystalline silicon (c-Si, mc-Si) and make up 85 % of the commercial solar cell market. The rest is shared by solar cells of the second generation, the so-called thin film solar cells, mainly based on amorphous or nanocrystalline silicon (a-Si, nc-Si), cadmium telluride (CdTe) and copper (indium, gallium) diselenide ($Cu(InGa)Se_2$) [1].

As silicon is after oxygen the most abundant element on earth and its crystalline solar cells have a high efficiency and long term stability, it is expected to dominate the solar energy marked also in future. For many niche applications instead, there is need for alternatives with reduced manufacturing costs and weight, increased flexibility or color adjustment, which is not only relevant for design aspects, but also for different illumination conditions, like indoors.

P. L. Marek, *Biomimetic Dye Aggregate Solar Cells*,
Springer Theses, DOI: 10.1007/978-3-319-00636-9_5,

The solar cells of the third generation aim to exceed the theoretical efficiency limit of about 30 %, the so-called Shockley–Queisser limit, which is the thermodynamic limit for conventional solar cells with one bandgap under not concentrated light [1, 2].

5.2.1 Silicon Solar Cells

As silicon can be doped to p- as well as n-type it allows the formation of so-called p–n-homojunctions within a single crystal. Due to this homojunction all essential solar cell functions, such as light absorption, charge separation and transport to different electrodes, are combined within one bulk material. This gives crystalline silicon a multifunctionality, which allows for a very high efficiency and good long term stability. This multifunctionality has also limitations making it less suitable for certain niche applications, because certain material properties can not be optimized individually. To achieve a high photon to current conversion efficiency silicon needs to be highly pure and crystalline, what makes the fabrication difficult and expensive. The low charge carrier recombination, which is a result of its indirect bandgap, causes on the other hand a low absorption coefficient, which requires relatively thick layers to absorb the full sunlight intensity. These layers are very brittle and need a stiff protection, like a thick glass pane, which causes a relatively high weight and consequently high costs. Furthermore, the efficiency drops significantly at low light intensities, making such solar cells less efficient for indoor applications, like for most consumer electronics. Hence, the energy, which is spend for manufacturing the solar cell might not get payed back during the lifetime of the device. These disadvantages led to a high effort in research for alternatives, like thin film solar cells or more recently solar cells based on organic semiconductors. For rooftop applications instead, the disadvantages of silicon solar cells are less relevant and their efficiency and long-term stability are more important.

Solar cells based on monocrystalline silicon reach efficiencies up to 25 % in the laboratory [3]. The highest efficiencies of commercially available solar cells are reached with monocrystalline silicon (c-Si), where the records are achieved, when both contacts are on the backside. In this so-called interdigitated back contact technique, electrons and holes are collected at the backside within two interdigitated metallic grids, which contact two differently doped regions. Thus, the front side is not partly shaded by a metal grid of the conventional back and front side contact design [4]. These cells reach with about 22 % the highest efficiencies in mass production and with 19.3 % the highest panel efficiencies, up to date [5]. Another strategy to improve the efficiency of conventional monocrystalline silicon solar cells is the so-called HIT type, which stands for heterojunction with intrinsic thin layer. In these cells, the conventional homojunction of a p- and n-doped region of the silicon single-crystal is replaced by the heterojunction between c-Si and amorphous silicon (a-Si). HIT solar cells reach already 20 % efficiency, and a panel efficiency of 17.3 % [6].

5.2.2 Thin Film Solar Cells

Thin film solar cells are one alternative, which already captured the market with various inorganic semiconductor heterojunction solar cells. They are based on a compound semiconductor with a direct band gap, which allows the full sunlight absorption within layers around 100 times thinner than those needed for crystalline silicon (c-Si). The most efficient absorbers are based on elements with the following valence combinations: III-V, II-VI, or II-III-VI_2. Their small thickness of a few μm allows also the fabrication of flexible solar cells on thermally resistant plastic substrates like polyimides or on metal foils. Thin film solar cells based on gallium arsenide (GaAs) show a record efficiency of 26 % in the laboratory, and thus the highest efficiency of all solar cells based on one band gap, so far. However, this efficiency is only possible with a single crystalline material, making it very expensive and limiting the size (lateral dimensions). The other materials mentioned above, reach relatively high efficiencies also as multicrystalline thin films and thus allow much lower manufacturing costs and larger cells.

Thin film solar cells based on multicrystalline copper (indium, gallium) diselenide $Cu(In,Ga)Se_2$ (CIGS) reach with 19.4 % the highest efficiency of multicrystalline thin film solar cells to date in the laboratory [3, 7]. Closely followed by multicrystalline cadmium telluride (CdTe), which achieved already 16.7 % efficiency in the laboratory [3]. Flexible CIGS solar cells printed on metal foils already reached an efficiency of 15.3 % [8]. Flexible CIGS solar cells on the lower temperature plastic substrate polyimide, fabricated by the economic roll-to-roll process reached already 13.4 % [9]. These efficiency records of flexible solar cells are closely followed by CdTe solar cells, which reach 12.4 % on the flexible polyimide foils [10].

The reduction of the manufacturing costs of such thin film solar cells is limited due to limited resources, which may counteract a cost reduction aspired by mass production. The deposition techniques needed for proper inorganic thin films require vacuum or inert atmospheres and elevated temperatures. Additionally, the toxicity of many inorganic thin film materials is undesired for low cost products as this would hamper the export to countries with a poor infrastructure for recycling.

Alternatively, the application of light-scattering and -trapping surface structures has been tested for a reduction of the thickness of crystalline silicon. However, the thickness reduction achieved so far is not enough to increase the bending properties significantly and such solar cells still remain brittle. Flexible thin film solar cells based on silicon are possible only with amorphous silicon (a-Si). There the indirect bandgap of crystalline silicon converts into a direct one, allowing also the full sunlight absorption within layer thicknesses in the order of a few micrometers. Such solar cells have the advantage of being based on the inexhaustible and environmentally friendly material Si, but suffer from a lower efficiency and lifetime compared to the crystalline inorganic thin film solar cells. Recently, a-Si solar cells reached also the 10 % efficiency mark [11].

5.3 Multijunction and Concentrator Cells

For multijunction solar cells, which are in principle a stacking of semitransparent solar cells, that absorb different wavelength ranges, the theoretical thermodynamic efficiency limit lies at 43 % for two junctions, that is, band gaps, 49 % for three and 66 % for infinity many gaps. As such solar cells are much more expensive, they are designed to work with concentrated sunlight to save material. Additionally, the theoretical efficiency limit is higher with concentrated sunlight, due to a higher contribution of multi-photon processes [1].

With concentrated light the theoretical Shockley–Queisser limit increases to 41 % for a single junction cell [2]. With the combination of a triple junction solar cell together with concentrated sunlight practical efficiencies of 41.1 % were already achieved by using $Ga_{0.35}In_{0.65}P/Ga_{0.83}In_{0.17}As/Ge$ on 5.09 mm^2 and a 454 fold sunlight concentration [12], whereas the latest confirmed record of multijunction concentrator solar cells is 41.6 % [13].

Such solar cells, however, need an active tracking of the lenses following the direction of the sun, what makes them only attractive for immobile applications in regions with a small fraction of diffuse sunlight.

5.4 Hybrid and Organic Solar Cells

Organic solar cells have a large potential for lowering the manufacturing costs, for example, due to their lower melting points, thus processing temperature, and the higher absorption coefficients compared to inorganic semiconductors. This enables an active layer of around 100-nm thickness, which is over 10 times thinner than that of inorganic thin films for solar cells. Their disadvantages, like a lower efficiency and shorter lifetime, commonly a few years, are of less importance for certain niche applications than the reduced fabrication costs, lightweight, high mechanical flexibility or a specialization for certain illumination conditions or design aspects. It might also be accepted to exchange the solar cell together with the rechargeable battery, which commonly has a shorter lifetime, than the device powered by it.

Due to the shorter exciton diffusion length in organic materials, which by far does not reach the approx. 100-nm dye thickness needed for full sunlight absorption [14], such solar cells are based on a mesoscopic interpenetrating network of donor and acceptor domains. Ideally, these domains are bicontinuous and have diameters in the order of twice the exciton diffusion length [15]. In this so-called bulk heterojunction, the excitons are able to diffuse to the charge separating interface where the donor injects the excited electrons to the acceptor from where those can reach their charge collecting electrode through the percolation path. Beside the light absorption the donor also transports the holes to the counter electrode.

Two concepts for the application of the donor–acceptor blends evolved. In one, small molecules are co-evaporated onto the substrate, like zinc phthalocyanines (ZnPc) as donors and C_{60}-fullerenes as electron acceptors. The other concept to form the bicontineous network of donor and acceptor domains is to dissolve these materials and apply them simultaneously by spin-coating or printing. The materials used therefore are commonly conjugated polymers as donors, like the poly(3-hexylthiophene) (P3HT) and fullerene derivatives with increased solubility as acceptors, like the phenyl-C_{61}-butyric acid methyl ester (PCBM).

This photovoltaic active blend is commonly applied onto a spin-coated electron blocking layer of poly(3,4-ethylenedioxythiophene): poly(styrenesulfonate) (PEDOT:PSS) on a transparent conductive substrate like indium-tin-oxide (ITO). The counter electrode is normally applied by evaporation of a thin calcium layer with an aluminum layer on top to prevent oxidation. These elements have the desired high work function, allowing for a high voltage output of the solar cell. Such non-noble metals however, are very reactive and require the evaporation at ultra high vacuum conditions and a good encapsulation. Recently, also TiO_2 has been introduced as hole blocking layer between the active blend and the aluminum counter electrode in organic bulk heterojunction solar cells, which achieved an efficiency of 6 % [16].

The highest efficiency of organic bulk heterojunction solar cells based on evaporated small molecules instead, was achieved in 2010 by the German company Heliatek GmbH with a tandem cell reaching 7.7 % efficiency [17].

The record efficiency for printable organic bulk heterojunction solar cells could be achieved in 2010 by the US company Solarmer Inc., reaching 8.13 % [18]. This high efficiency was possible by using a new donor polymer class, the poly(thieno[3,4-*b*]thiophene-benzodithiophene) (PTB), which has been developed in 2009 by Yongye Liang et al. [19] and which outperformed the formerly most common polymer dye P3HT. Together with the substitution of the C_{61} core of PCBM with C_{71} the efficiency could be increased by this research group to 7.4 %, due to a complimentary light absorption by this new donor–acceptor blend [15, 20].

Berson et al. reported that P3HT forms nanofibers in certain solvents with intermediate solvation potential, like p-xylene, causing the solution to gel at room temperature [21]. This allowed to improve the efficiency without the common thermal post-treatment. As these P3HT nanofibers have a red-shifted absorption spectrum [22], we can assume, that these polymers also self-assemble to the ordered structure of J-aggregates. This shows that self-assembly may be the reason for the high efficiency of these dyes.

Also carbon nanotubes have been investigated as replacement for the PCBM, but the efficiencies around 2 % were roughly half of those obtained with PCBM, most probably due to their lower solubility [21, 23].

5.5 Dye Sensitized Solar Cells

The highest efficiencies of solar cells with organic absorbers have been reached with the so-called DSSC, also-called dye solar cell (DSC) or *Grätzel cell*, named after its inventor Michael Grätzel, who developed it in 1991 together with Brian O'Regan [24]. They reached a confirmed efficiency of 11.2 % [25]. Another advantage is that their manufacturing does not require for inert gas or vacuum deposition techniques.

Common DSSCs are fine-porous colorless TiO_2 substrates stained with a monolayer of an organic or metalorganic dye. TiO_2 is an intrinsically n-doped semiconductor with a band gap, capable only for absorption of ultraviolet light lower than approx. 400 nm wavelength. To sensitize it for the lower energetic optical light, it needs the contact to a dye, which injects its excited electrons into the TiO_2 conduction band, if the energy levels matches in the right way. As this injection needs some time, in the order of picoseconds, an efficient dye needs the ability to conserve its excitation long enough. Thus, a non-radiative recombination of its excited electron–hole pairs, the so-called excitons needs to be prevented. This recombination is significantly enhanced by additional vibronic states, which are induced by disordered dye aggregation. As this recombination process is around 1000 times faster than the radiative recombination, it reduces significantly the sensitization efficiency. As the vast majority of organic dyes only form disordered aggregates, conventional dyes are commonly used in the form of monolayers covering the TiO_2 and not as multilayers or aggregates. Organic dyes have the highest absorption coefficients and are potentially cheap in fabrication, a fact that makes them attractive for low cost solar cells. To accommodate enough dye for full sunlight absorption onto the TiO_2 surface, keeping the pathway for charges short, a high surface area of TiO_2 within a small volume is needed. This requires so-called mesoporous structures of TiO_2 layers, which is achieved by sintering TiO_2-nanoparticles with sizes around 20 nm to layers of about. 10–14 μm thickness. Using commercially available TiO_2 nanopowders, efficiencies up to 9.2 % have been achieved [26]. The lower efficiency compared to cells made of home-made TiO_2 nanopowders is due to the higher amount of rutile (about 20 % rutile and 80 % anatase) in commercial TiO_2 nanopowders like the P25 from *Evonik*, which is fabricated in big quantities by fumed $TiCl_4$ synthesis.

The hole transportation in DSSCs is commonly carried out by iodide-based electrolytes, which are able to penetrate the tiny pores and contact each dye molecule. As the sealing of the volatile electrolyte is problematic and known as the main factor limiting the lifetime of these solar cells, also ionic liquids have been developed as substitute for the conventional electrolyte. Thus, the leakage and evaporation problem could be solved due to their much lower vapor pressure. Because of the higher viscosity, and thus higher ionic resistivity, the efficiency is slightly reduced, reaching 8.2 % up to date [27].

The liquid electrolyte, used for hole transportation, unfortunately does not match the goal of lightweight, mechanical flexibility or fast assembly. Thus, much effort was invested into the research for suitable solid substitutes for the liquid electrolyte.

5.5.1 Conventional Dyes for DSSCs

The best performing dyes in terms of efficiency as well as stability are ruthenium polypyridin complexes, like the N719 dye. They reach over 11 % efficiency in conventional DSSCs (Fig. 5.1) [28–30].

Ruthenium polypyridine dyes inject excited electrons to the TiO_2 conduction band with a high yield of almost unity considering the losses due to reflections on the glass. The reason for this is an efficient intersystem crossing from the singlet to the triplet state, facilitated by the heavy metal element Ruthenium. Transient spectroscopy measurements on ruthenium and osmium polypyridyl complexes showed that the dominating pathway for injection of electrons into the TiO_2 conduction band is from the triplet state of the dye. Only a minor fraction gets injected directly from the singlet state [31]. Within the triplet state the lifetime of the excited electrons is much higher than in the singlet, which allows for the injection, until recombination to the ground state gets competitive.

In the field of organic light emitting devices (OLED) the implementation of heavy metal atoms, like iridium, into the luminescent molecules, leads also to the most efficient devices [32]. In this case, the heavy element facilitates the intersystem

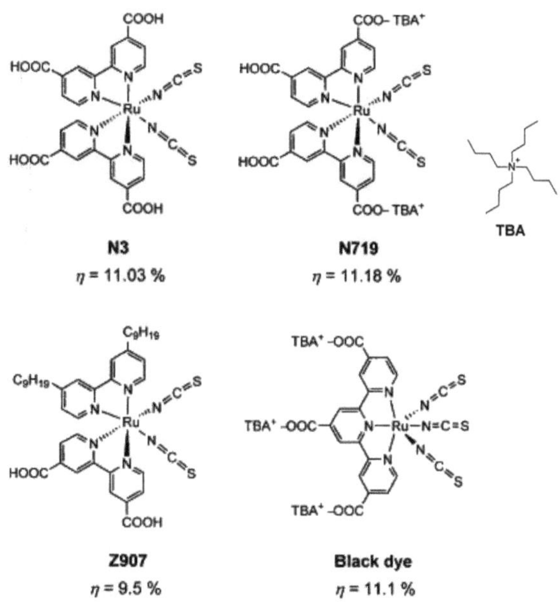

Fig. 5.1 Structures of conventional ruthenium polypyridine dyes (Ru-dyes). These are the structures of the most efficient ruthenium polypyridine dyes with their efficiencies achieved in DSSCs [29], TBA: Tetrabutylammonium cation. Copyright (2009) Wiley–VCH Verlag GmbH & Co. KGaA, Weinheim

crossing from the triplet state to the singlet state, from where a radiative recombination is possible. These compounds are therefore called triplet emitters.

Fig. 5.1 shows the structures of some of the most efficient ruthenium polypyridine dyes with their efficiencies achieved in DSSCs.

Many more ruthenium polypyridin dyes have been synthesized so far, which have also different colors. The *Sony* Corporation produced some multicolored DSSC-prototypes demonstrating their potential for commercial applications, as could be seen at the exhibition *Eco-Products* 2008 in Japan (Fig. 5.2) [33, 34].

The energetics and kinetics of excited electrons at the dye-TiO_2 interface for the most common ruthenium dye N719 are explained in detail by Koops et al. [35]. The main pathway for the excited electron injection is from the triplet state of the dye. The transfer from the singlet to the triplet state occurs within about 100 fs. The formally forbidden recombination within the triplet state occurs on a time scale of 10–100 ns, which is 100–1000 times slower than the electron injection into the TiO_2 (10–100 ps) [36]. The regeneration of the dye by the iodide electrolyte occurs at a half-time of approx. 1 μs, fast enough to compete with the recombination of electrons from the TiO_2 conduction band with holes in the dye ground state, which is in the timescale of μs to ms. The recombination of electrons from the TiO_2 with holes from the electrolyte is in the order ms to s [36].

Recently, an efficiency of 9 % has been reached even with a metal-free organic dye, which promises lower manufacturing costs due to savings in resources [29, 37]. Ruthenium-based dyes instead, have a higher stability and may keep 96 % of the initial efficiency (over 9 %) at standard AM1,5 irradiance and 60 °C thermal stress after 1000 h [38].

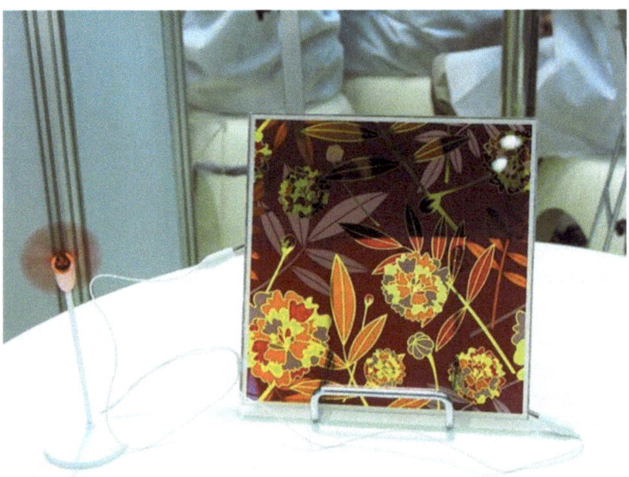

Fig. 5.2 Photo of a multicolored DSSC. This commercial semitransparent and multicolored DSSC from the *Sony* Corporation is powering a small vent on the exhibition Eco-Products 2008 in Japan [33, 34]. Reproduced with Permission from DIGITIMES Inc. Copyright (2008)

5.5.2 *Natural Dyes*

The photosensitization of a wide-bandgap semiconductor (ZnO) by chlorophylls was first investigated by Tributsch and Calvin in 1971 but with very low photocurrents although quantum efficiencies of 0,125 electrons per absorbed photon were achieved [39].

Kay and Grätzel investigated in 1993 several derivatives of chlorophylls and related natural porphyrins as sensitizers in DSSCs. They found that free carboxylic groups were important for the adsorption onto TiO_2, and they reached a conversion efficiency of 2.6 % on the 12 μm thick mesoporous TiO_2 film, which was about 1/3 compared to that of conventional ruthenium dyes [39]. Wang et al. achieved with chemically modified chlorophyll derivatives containing carboxylic groups and the classical iodide-based electrolyte a conversion efficiency of 3.1 %, which could be increase to 4.0 % by the addition of bacterial carotenoids [40]. The efficiencies of chlorophyll derivatives equipped artificially with carboxylic binding groups could be further improved to 6.5 % [41, 42].

Also potentially cheap and abundant natural dyes have been tested in such solar cell architectures, like an anthocyanin dye, extracted from blackberries, which reached a conversion efficiency of 1 % [43]. Other natural dyes reached up to 1.7 % efficiency sensitizing TiO_2 as a monolayer, which was stated to be among the highest efficiencies achieved with raw natural dyes [44].

5.5.3 *Porphyrin Sensitizers*

Zn–porphyrin sensitizers are among the best performing porphyrin-based dyes for DSSCs [45]. Whereas efficiencies with zinc-porphyrin monolayers and the common iodide-based electrolyte reached 7.1 %, a new electrolyte based on a Co(II/III)tris(bipyridyl) redox couple, allowed recently also efficiencies up to 12.3 % for a new zinc porphyrin sensitizer [46]. In the all-solid state version with spiro-MeOTAD as hole transport material 3.6 % efficiency could be reached, what is quite close to the best performing ruthenium-based dyes in solid-state DSSCs [45].

Mozer et al. investigated the typically encountered lower open circuit voltages of porphyrin (monolayer) sensitized solar cells of about 100–200 mV compared to that sensitized by Ruthenium-dyes [47]. These authors attributed this to an increased recombination rate, which may originate by their planar molecular geometry compared to the more spherical one of Ru-complexes. The planar geometry enables the porphyrins to adsorb flat lying on the TiO_2 surface and attract the negatively charged iodide ions by their positively charged center. This may enhance the recombination rate compared to that with Ru-dyes, which have negatively charged NCS ligands repulsing the iodide anions.

5.5.4 Dye Aggregates in DSSCs

One of the first rare attempts of using dye aggregates in DSSCs instead of dye monolayers was done in 1996 by Nüesch et al. [48]. The authors used a self-assembling merocyanine dye which was supposed to form J-aggregates within the mesopores of TiO_2, Al_2O_3, and ZrO_2 layers. The J-aggregates functioned as sensitizers, but were difficult to control due to the humidity in the air, which facilitated the formation of H-aggregates. In the mesoporous Al_2O_3 and ZrO_2 layers no J-aggregate formation could be induced, which was attributed to not sufficiently large areas of atomically flat surfaces within the mesopores, which were supposed to induce the growth of J-aggregates until they reach a stable size.

As the number of dyes with the ability to self-assemble to highly ordered structures is very small, the focus in the DSSC research lies rather in prevention of the aggregation of conventional dyes than in the improvement of the aggregate growth mechanism. Disordered aggregates have a high exciton quenching rate due vibrational interactions. This effect can be already observed, when a solution of fluorescent monomers becomes highly concentrated. The high rate of random contacts between the monomers leads to the so-called concentration quenching of the fluorescence. This radiationless recombination is orders of magnitudes faster than a radiative recombination. Beside ruthenium polypyridine dyes, also conventional porphyrin and phthalocyanine dyes, used as monolayers in DSSCs, are strongly affected by disordered aggregation [49]. The disorder within one monolayer may be already a big problem for DSSCs, especially for solid-state devices as will be shown in the next chapter.

Instead of struggling against this aggregation it might be possible to learn from nature how to force aggregation in a sophisticated way and make use of self-assembly to achieve highly ordered aggregates, which do not disturb solar cell functionality, but rather improve it.

5.6 Solid-State Dye Sensitized Solar Cells

In a solid-state dye-sensitized solar cell (ss-DSSC), also sometimes abbreviated to: SSDSC [50] or SDSC [51], the liquid electrolyte for the hole transport is replaced by a solid hole transport material (HTM). The HTM is intended to solve the technological problems encountered with the handling and sealing of the liquid electrolyte, reducing the solar cell lifetime, due to degradation of the sealant and drying-out of the cell [52]. Additionally, the solid HTM would in principle allow the manufacturing of lighter and flexible solar cells in the cost efficient roll-to-roll process.

However, the efficiencies of ss-DSSCs are generally much lower than those based on the liquid iodide electrolyte. A breakthrough was achieved by Bach et al. in 2000 with the investigation of the HTM spiro-MeOTAD in combination with certain additives and dopants. These authors could rise the efficiency from 0.04 % without additives to 0.75 % with the addition of a chemical dopant, known as antimony dopant and the ionic dopant TFSI-Li, which will be described later in more detail. Further improvements could be achieved since this time by optimizing the additive composition and the implementation of light scattering layers to reach efficiencies between 4-5 %.

5.6.1 Improvement of ss-DSSCs

The two main reasons for the lower performance of ss-DSSCs compared to the liquid electrolytes are: (1) a shorter electron diffusion length due to a higher recombination rate at the HTM/TiO_2 interface and (2) an insufficient filling of the mesopores in the TiO_2 network [36].

The charge carrier diffusion length within the spiro-MeOTAD filled TiO_2 layers is only a few micrometers, compared to 20-100 μm in layers with the liquid iodide-based electrolyte. The recombination rate in solar cells with spiro-OMeTAD is almost two orders of magnitude larger than in those with the liquid electrolyte [53]. One reason for this higher recombination rate between TiO_2 and solid HTMs is that the recombination is a one-electron-process, compared to the two electron process of the recombination between TiO_2 and the iodide electrolyte [54]. The one-electron recombination process is much more probable and faster. Additionally, the iodide ions are negatively charged and may get repelled from the TiO_2 surface, which is also negatively charged. The spiro-MeOTAD instead is positively charged, which may facilitate the recombination [55].

The second major problem is the infiltration of the mesoporous TiO_2 layers, which becomes worse with increasing the TiO_2 layer thickness.

Both effects cause the efficiency maximum of ss-DSSCs to be at much lower mesoporous TiO_2 layer thicknesses, than with the liquid hole transportation. The optimum TiO_2 layer thickness for ss-DSSCs is about 2 μm, whereas that for DSSCs with the liquid iodide-based electrolyte is 10–14 μm. The reduced light absorption within such thin layers limits the overall efficiency, which can be obtained.

Using special vacuum techniques aiding the infiltration of the solid HTM, the pore-filling and thus the applicable TiO_2 layer thickness can be increased [56]. However, these techniques are not well established yet, maybe because the improvement is not sufficient, as the pore-filling is mainly determined by the solubility of the HTM. Soluble polymers can not be infiltrated and need to be polymerized within the mesopores, which may cause damages in the dye coating. Another possibility to increase the light absorption in thinner layers is to increase the retention time of the light by the implementation of light scattering structures.

A third reason for the efficiency decrease in ss-DSSCs, besides the enhanced recombination at the TiO_2/spiro-MeOTAD interface and the worse pore-filling with spiro-MeOTAD may be an enhanced exciton quenching within the dye monolayer. Schwanitz et al. showed by photoelectron spectroscopy (PES) measurements on Ru-dye monolayers on TiO_2, that the ordering within the monolayer decreases in the absence of the standard electrolyte solvent acetonitrile (AN) [57]. An overlap of the dye molecules within the monolayer may quench the excitons. The solvent in DSSCs is supposed to penetrate the dye monolayer and prevent direct contacts between the dye molecules. In this PES measurement Schwanitz et al. measured in the absence of AN a broadening of the photoelectron spectrum, an indication for a random overlap of orbitals. The graphics in Fig. 5.3 are extracted from this work and show the PES measurement of such a Ru-dye monolayer with and without AN and schematically the possible alignment of the dye monolayer interacting with the solvent molecules.

Self-assembling dyes, forming well-defined multilayers or aggregates with the ability to transport excitons, could improve conventional ss-DSSCs for the following reasons: (a) they do not form disordered intermolecular contacts, which lead to fast exciton quenching, (b) they can be packed onto a smaller TiO_2 area, which is supposed to be highly affected by recombination with the HTM, (c) the TiO_2 layer could be coarsened, improving its infiltration with various HTMs, and (d) a coarse-porous TiO_2 layer would have a higher light scattering effect, increasing the light absorption.

In the following part commonly used materials for ss-DSSCs will be described shortly.

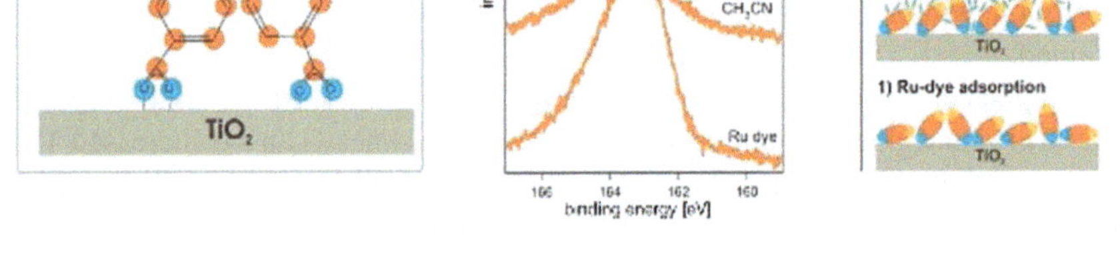

Fig. 5.3 The solvent effect of acetonitrile on the aggregation of a Ru-dye monolayer. These graphics form Schwanitz et al. [57] show a model of the standard ruthenium bipyridyl dye N3 on a TiO_2 surface (**a**), four photoelectron spectroscopy measurements of a monolayer of this dye on TiO_2 before co-adsorption of acetonitrile (AN), after co-adsorption of three monolayers of AN, after additional three monolayers of AN and after the desorption of AN (**b**). The broadening of the spectrum without the co-adsorbed solvent molecules is an indication for an increase in disorder of the monolayer, which is schematically depicted in (**c**). Copyright (2007) Schweizerische Chemische Gesellschaft

5.6.2 *Electrode Materials for ss-DSSCs*

5.6.2.1 FTO

As transparent electrode the fluorine doped tin oxide (SnO_2:F), abbreviated to FTO, is commonly used for DSSCs and ss-DSSCs. It is commercially available as thin layer (about 550 nm) on glass substrates and allows the highest efficiencies of these solar cells although indium tin oxide (In_2O_3:Sn, ITO), which is commonly used for organic bulk heterojunction solar cells, has a better transparency. The reason why FTO performs better than ITO in DSSCs may be its higher stability at elevated temperatures around 450 °C, needed for sintering the TiO_2 layer. Additionally, its work function of 4.4 eV does better match the conduction band edge of TiO_2 (4.2 eV) compared to the work function of ITO, being around 4.7 eV. In contrast to FTO the work function of ITO depends strongly on the cleaning methods, thus can vary between 4.4 and 4.8 eV [58].

5.6.2.2 TiO_2 Nanoparticles

As anatase, the low temperature modification of TiO_2, is known to perform better in DSSCs than rutile, practically only anatase nanoparticles are used for these solar cells. Park et al. showed, that the efficiency of DSSCs made of rutile nanoparticles is about 30 % less than with anatase particles [59]. These authors could relate the deficit mainly to the smaller surface area of the slightly larger rutile particles, and hence to the lower dye amount adsorbed to it. A good description how to produce the right TiO_2 nanoparticles and assemble DSSCs, reaching efficiencies over 10 % is given by Ito et al. [60]

5.6.2.3 TiO_2 Hole Blocking Layer

As in DSSCs the liquid electrolyte already forms a current rectifying barrier with the FTO, a compact TiO_2 hole blocking layer on the FTO underneath the mesoporous TiO_2 layer is not essential to obtain a reasonable device performance. For ss-DSSCs instead, Peng et al. showed, that a hole blocking layer is an indispensable component, as scarcely any output power could be obtained for these cells without any blocking layer on the FTO. The short circuit current (I_{SC}) was below 10^{-7} A/cm^2 and the open circuit potential (V_{OC}) was less than 1 mV [61]. This is due to the formation of an ohmic contact between spiro-MeOTAD and the FTO. Nowadays such blocking layers are used for DSSCs with liquid electrolytes, too, in order to further improve the efficiency [62], especially at low light intensities [63] and for organic dyes, which benefit, in contrast to ruthenium polypyridine dyes, significantly by the blocking layer. The efficiency of DSSCs with organic dyes and a hole blocking layer increased by a factor of 1.6 at full light intensity and up to a factor of 2 at low light intensities [64] compared to cells without the hole blocking layer. The photovoltage dropped significantly at incident photon fluxes below $10^{17}/(cm^2\ s)$ without a blocking layer, whereas with a blocking layer, the slope of the photovoltage decrease remained constant, even at five orders of magnitude lower light intensities.

To avoid any direct contact of the FTO with the spiro-MeOTAD, an approx. 100 nm thick compact TiO_2 layer is applied onto the FTO prior to the application of the mesoporous TiO_2 layer. The compact TiO_2 layer is commonly applied by the spray pyrolysis deposition (SPD) of di-iso-propoxy titanium bis(acetylacetonate), denoted to TAA, onto a 450 °C hot FTO substrate [65].

The efficiency can be further improved, when the oxygen vacancy concentration in the compact TiO_2 layer is minimized by using oxygen as carrier gas for the spray pyrolysis deposition [66, 67]. To avoid a high oxygen vacancy concentration, which leads to a barrier formation at the FTO/TiO_2 interface due to the Fermi-level rise of n-doped TiO_2, the storage of the samples with readily prepared TiO_2 hole-blocking layers under vacuum or inert gas should be also avoided, as oxygen tends to diffuse out of the TiO_2 under these conditions.

In this thesis, the hole blocking layers of compact TiO_2 have been applied by the atomic layer deposition technique onto FTO with subsequent post-annealing in order to increase the oxygen occupancy.

5.6.2.4 Porous TiO_2 Layer

Mesoporous TiO_2 layers are commonly produced by doctor blading or screen printing dispersions of TiO_2 nanoparticles in ethanol, water or terpineol, containing also organic binders like ethyl cellulose to allow thicker crack-free layers. After sintering these layers at 450–500 °C, their performance can be improved by a treatment within aqueous solutions of $TiCl_4$, which cause the growth of an additional 1–2 nm thick TiO_2 layer all over the mesoporous structure [60, 68]. This additional TiO_2 layer improves the photocurrent due to a decrease in the recombination rate constant [69]. Also the dye uptake and electron injection rate are increased by the $TiCl_4$ treatment. More details about the application of high-performance mesoporous TiO_2 layers are given in the precise description from Ito et al. [60]. As iron ions do quench efficiently the generated photocurrent by increasing the recombination rate [60], anything containing iron was avoided during the cell preparation. The pore volume of typical mesoporous TiO_2 layers is about 60 % [70].

5.6.2.5 Light Scattering Layer

Light scattering layers increase the retention of the light within the layer and thus help to reduce the mesoporous TiO_2 layer thickness and the amount of dye needed for full sunlight absorption. Additionally, shorter pathways through thinner layers reduce the internal electrical resistance.

Typically, anatase particles within the size range from 300 to 400 nm are applied as a scattering layer on top of a mesoporous layer, but these particles may also be implemented as one mixed layer of nano- and microparticles.

Pure light scattering layers with only 20 % addition of 10-nm sized nanoparticles in order to improve the sintering at lower temperatures, do also significantly contribute to the photocurrent generation. Zhang et al. achieved with such 5-μm thick coarse-porous TiO_2 layers (with 20 % nanoparticles) impressive 5 % overall efficiency in DSSCs with the common iodide-based electrolyte, although, their surface and thus dye uptake is much smaller than that of mesoporous layers. The reason for the relatively high efficiency without a mesoporous layer is supposed to be, beside the efficient light trapping within the layer, also a two-fold reduced density of trap states on the TiO_2 surface [71].

5.6.2.6 Counter Electrode

As counter electrode, commonly a 30–100-nm thick gold film is evaporated on top of the spiro-MeOTAD layer. These solar cells benefit from a metal electrode, as it reflects the transmitted light back, increasing the effective path way through the cell. Although the reflectivity of gold below 450 nm decreases to ca. 30 %, it is the most used metal for ss-DSSCs, due to its close matching the HOMO level of spiro-MeOTAD (work function gold: 4.9 eV [36], HOMO spiro-MeOTAD: 5.1 eV [72]).

Snaith et al. reported, that counter electrodes of silver increase the efficiency of ss-DSSCs compared to gold [73], but silver electrodes are known to be less stable than gold ones as they tend to oxidize.

5.6.3 Hole Transport Material

The first successful implementation of a solid hole transport material (HTM) into DSSCs was demonstrated 1998 by Bach et al. [74]. These cells consisted of spiro-MeOTAD, also called spiro-OMeTAD [75], as substitute for the liquid electrolyte for the hole transport. It is the p-type triarylamine derivative: 2,2',7,7'-tetrakis(*N*,*N*-di-*p*-methoxyphenyl-amine)-9,9'-spirobifluorene [76] (Fig. 5.4).

Bach et al. first used it together with a hole blocking layer of TiO_2, applied by the spray pyrolysis deposition technique, the chemical p-dopant: $[N(p\text{-}C_6H_4Br)_3][SbCl_6]$, also called Sb-dopant, and the ionic dopant: $Li[CF_3SO_2]_2N$, also called TFSI-Li. An efficiency of 0.74 % was achieved, whereas without the additives the efficiency was only 0.04 %. Since this time the efficiency could be increased to over 4 % mainly by adjusting the additives for the spiro-MeOTAD, which is up to date still the best performing HTM for ss-DSSCs [75]. The Sb-dopant is not in use anymore, but the 4-tert-butylpyridine (tBP) is an essential additive for these solar cells by now.

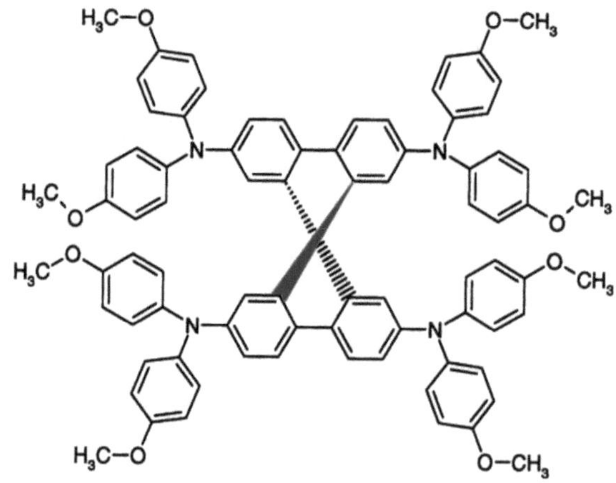

Fig. 5.4 Structural formula of spiro-MeOTAD. The wedged and hatch-wedged bonds in the center of the spiro-MeOTAD formula, adopted from [76], indicate the twist of the northern and southern part, which are oriented perpendicular to each other. Copyright (2005) American Institute of Physics

Spiro-MeOTAD is normally processed by spin coating from chlorobenzene (CB) solutions. The hole mobility of spiro-MeOTAD is not the highest of potential HTMs [77], but its wetting of TiO_2 together with its relatively small molecular size results in a good TiO_2 mesopore infiltration, which makes it to the best performing HTM in ss-DSSCs reaching efficiencies around 4-5 % [52, 77, 79–81]. The main limitations of such solar cells are attributed to a high recombination rate at the interface between spiro-MeOTAD and TiO_2 and a high hole transport resistance in the spiro-MeOTAD, which leads to a relative low fill factor (FF) [53].

In the following paragraph the additives for the spiro-MeOTAD matrix will be shortly described. As the additives and their concentrations have been changed and varied a lot since the first usage of spiro-MeOTAD for ss-DSSCs and many other parameters additionally affect the solar cell efficiencies, there is no real consensus about the ideal additive composition. The table in the appendix gives a small overview about various materials, additives and other parameters, for example, thickness and sintering temperature of the mesoporous TiO_2 layer, which have been reported for the construction of ss-DSSCs and DSSCs.

5.6.3.1 TFSI-Li

Snaith et al. [78] showed that the addition of Bis(trifluoromethane)sulfonimide lithium salt (TFSI-Li) significantly improves the hole mobility of spiro-MeOTAD. Upon the addition of 12 % Li-salt into a 2.2 μm thick spiro-MeOTAD layer sandwiched between two gold electrodes the mobility of holes increased from $1.6 \cdot 10^{-4}$ to $1.6 \cdot 10^{-3}$ $cm^2/(Vs)$. By Kelvin probe microscopy with channel length variation it could also be shown, that the conductivity increase due to the addition of TFSI-Li is a bulk effect and not caused by an improved charge injection from the gold electrodes, what is known to be the operational mechanism of such ions in organic light emitting diodes. It was assumed that this ionic dopant reduces the potential barrier between neighboring spiro-MeOTAD anions and thus increases the hole mobility. Calculations predicted that 12 mol- % TFSI-Li would lead to an anion–anion separation of 2.6 nm and thus reduction of the barrier height of the potential well with the radius of 19 nm by up to 0.7 eV. Another explanation for the increase of hole mobility may arise from the increase of the polarizability of the matrix, which may result in a reduction of the binding energy of polarons and thus reduction of the effective mass of holes.

5.6.3.2 Sb-dopant

Although the chemical p-dopant (tris(p-bromophenyl)ammoniumyl hexachloroantimonate $[N(p\text{-}C_6H_4Br)_3][SbCl_6]$) is still used sometimes it is not recommended anymore as dopant for spiro-MeOTAD, as it does not enhance the performance in combination with the better working TFSI-Li. It increases the conductivity of spiro-MeOTAD, but this is counterbalanced by the increased recombination rate [78].

5.6.3.3 tBP

Krüger et al. reported that the additive 4-*tert*-butylpyridine (tBP) in combination with TFSI-Li lowers the recombination rate at the interface between TiO_2 and spiro-MeOTAD and increases the open circuit voltage of these solar cells. Additionally, it increases the solubility of lithium ions within the spiro-MeOTAD solution, due to its higher polarity, and makes the films smoother. Hence, it is assumed to increase also the wetting of TiO_2, and thus facilitates the infiltration of mesoporous TiO_2 films with the spiro-MeOTAD matrix [82]. Howie et al. investigated the potential shift toward higher open circuit voltages of about 0.25 V by the addition of tBP and could confirm the assumptions by electrochemical impedance spectroscopy measurements, that this is related to a shift of the flat band potential of the TiO_2 toward about 300 mV higher negative potentials. Additionally, the doping density of the layer increased by the addition of tBP from $2 \cdot 10^{18}$ to $7 \cdot 10^{18}$ cm^{-3} [83].

For porphyrin (monolayer) sensitized solar cells 4-tBP has the same positive effect of increasing open circuit voltage (V_{OC}) as for the Ru-dye sensitized ones. The V_{OC} of porphyrin sensitized solar cells, which is approx. 100–200 mV lower than that of their Ru-dye analogues, drops further by up to 200 mV in the absence of 4-tBP in the electrolyte.

5.7 Solar Cell Characterization

In order to investigate the quality of a solar cell and determine its efficiency measurements of the current–voltage characteristics (I-V curve) are required, which can be measured with a source measuring unit or a potentiostat. The efficiency of a solar cell is determined by the ratio of the maximum electrical power, extracted at the maximum power point (MPP) in the I-V curve, and the radiant power of the incident light:

$$\eta = \frac{V_{\mathrm{MPP}} \cdot I_{\mathrm{MPP}}}{P_{\mathrm{light}}}$$

where η is the efficiency of the solar cell, V_{MPP} and I_{MPP} are the potential and current at the MPP, respectively, and P_{light} is the incident light power.

The I-V curve reveals also other parameters, which can help to optimize solar cells, like the series and parallel (shunt) resistance and the so-called FF. The FF is the ratio of the maximum power output and the product of the open circuit voltage (V_{OC}) and the short circuit current (I_{SC}):

$$\mathrm{FF} = \frac{V_{\mathrm{MPP}} \cdot I_{\mathrm{MPP}}}{V_{\mathrm{OC}} \cdot I_{\mathrm{SC}}}$$

The FF can be visualized in an I-V curve as the ratio between two rectangular areas as shown in the fourth quadrant in Fig. 5.5 [84].

The efficiency of a solar cell is measured at standardized illumination conditions at 25 °C and simulated sunlight with a radiant power of 100 mW/cm^2. The spectrum corresponds to that of sunlight after passing 1.5 times the air mass (AM), which would be passed at a perpendicular irradiance at the equator. The solar spectrum of this so-called air mass 1.5 approximates that of a clear day at noon time in the beginning and end of summer at the central European latitude.

In order to investigate the solar cell performance also at indoor illumination conditions a measurement setup was built up with the following illumination

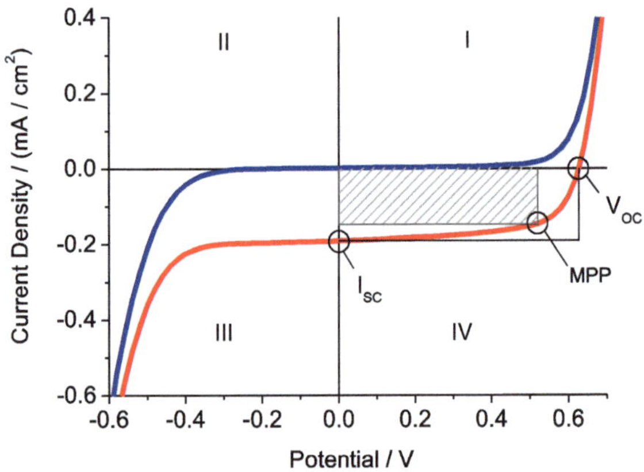

Fig. 5.5 I-V curve: example of a solar cell characteristic. This I-V curve, measured in the dark (*blue*) and under illumination of the lamp (60-W incandescent lamp) (*red*), shows the typical characteristic of a solar cell on the example of a DSSC with the ruthenium dye N719 and an iodide-based electrolyte. The maximum power output of the solar cell is highlighted as shaded area in the quadrant IV. Scan rate: 0.5 V/s, step potential: 0.02 V

conditions. The radiation source was an incandescent lamp with a reflecting metallic coating at the rear side and a power of 60 W, which is in the following denoted as "the lamp." The solar cell samples were mounted in a fixed position in 10 cm distance to the bulb surface of the lamp, as shown in Fig. 5.5. Some selected solar cells were measured additionally under the irradiation with simulated sunlight in cooperation with the group of Prof. W. Jaegermann at the Technical University of Darmstadt and the group of Prof. U. Lemmer at the Karlsruhe Institute of Technology.

Fig. 5.5 shows a typical characteristic of a DSSC with the conventional ruthenium dye N719, in the dark and under irradiance with the lamp, which is shown in Fig. 5.6.

The diode characteristic of the I-V curve measured in the dark (blue trace) shifts to negative photocurrents upon illumination (red trace). The almost horizontal slope is inversely proportional to the shunt resistance and the steep slope in the first and third quadrant is inversely proportional to the series resistance. An increase of the shunt resistance and decrease of the series resistance increases the fill-factor, and hence the efficiency.

For this DSSC, we added SiO_2 nanoparticles to the electrolyte (*Iodolyte AN-50* from *Solaronix*) in order to obtain a gel, and thus reduce the leakage rate of the electrolyte [85].

References

1. Crabtree GW, Lewis NS (2007) Phys Today 60:37–42
2. Shockley W, Queisser H (1961) J Appl Phys 32:510–519
3. Green MA, Emery K, Hishikawa Y, Warta W (2009) Prog Photovolt Res Appl 17:320–326
4. Characterization and Modeling of Back-Contact Back-Junction Silicon Solar Cells Kluska S, Granek F, Rüdiger M, Hermle M, Glunz SW 24th European Photovoltaic Solar Energy Conference 2009; Hamburg, Germany 950–954; http://www.ise.fraunhofer.de/de/veroeffentlichungen/konferenzbeitraege/2009/24th-europeanphotovoltaic-solar-energy-conference-and-exhibition-hamburg-germany/kluska_2ao.2.5.pdf
5. 315 Solar Panel description, Exceptional Efficiency And Performance, SunPower Corporation (05/2009) http://abcsolar.com/pdf/SunPower_318_Solar_Module.pdf
6. HIT photovoltaic module description / data sheet, HIT-240HDE4, SANYO Electric Co.,Ltd. (10/2009) http://www.cc-solar.gr/img/products/sanyo/hit1.pdf
7. Repins I, Contreras MA, Egaas B, De Hart C, Scharf J, Perkins CL, To B, Noufi R (2008) Prog Photovolt Res Appl 16:235–239
8. NREL Certifies 15.3% Nanosolar Foil Efficiency, Nanosolar (09/2009) http://cache.rmartinr.com/NanosolarCellWhitePaper.pdf
9. Weltrekord: 13,4 Prozent Wirkungsgrad bei Solarzellen auf Kunststofffolie, Solarion AG (10/2009) http://www.solarserver.com/solar-magazine/solar-news/current/german-pv-producer-solarionag-announces-134-conversion-efficiency-for-solar-cells-on-plastic-film.html
10. Flexible CdTe on polymer film achieves a new record efficiency of 12.4%, globalsolartechnology, (08/2009) http://www.pv-tech.org/news/flexible_cdte_thin-film_cells_reach_12.4_efficiency
11. World Record Efficiency for Amorphous Silicon Photovoltaic Solar Cells, Azocleantech (09/2009) http://www.azocleantech.com/details.asp?newsID=6675

12. Weltrekord: 41,1% Wirkungsgrad für Mehrfachsolarzellen am Fraunhofer ISE, ISE (01/2009) http://www.ise.fraunhofer.de/presse-und-medien/presseinformationen/presseinformationen-2009/weltrekord-41-1-wirkungsgrad-fuer-mehrfachsolarzellen-am-fraunhofer-ise
13. NREL confirms Spectrolab solar cell hits new conversion efficiency record of 41.6%, PV-tech (08/2009) http://www.pv-tech.org/news/_a/nrel_confirms_spectrolab_solar_cell_hits_new_conversion_efficiency_record_o/
14. Greenham NC, Grätzel M (2008) Nanotechnology 19:420201
15. Liang Y, Yu L (2010) Acc Chem Res 43:1227–1236
16. Park SH, Roy A, Beaupre S, Cho S, Coates N, Moon JS, Moses D, Leclerc M, Lee K, Heeger AJ (2009) Nat Photon 3:297–302
17. Heliatek and IAPP achieve record efficiency levels for organic solar cells, Heliatek GmbH / 2010
18. Solarmer Energy, Inc. Breaks Psychological Barrier with 8.13% OPV Efficiency, ad-hoc-news (07/2010) http://www.businesswire.com/news/home/20100727005484/en/Solarmer-Energy-Breaks-Psychological-Barrier-8.13-OPV
19. Liang Y, Feng D, Wu Y, Tsai S-T, Li G, Ray C, Yu L (2009) J Am Chem Soc 131:7792–7799
20. Liang Y, Xu Z, Xia J, Tsai S-T, Wu Y, Li G, Ray C, Yu L (2010) Adv Mater 22:E135–E138
21. Berson S, De Bettignies R, Bailly S, Guillerez S (2007) Adv Funct Mater 17:1377–1384
22. Berson S, Cecioni S, Billon M, Kervella Y, de Bettignies R, Bailly S, Guillerez S (2010) Sol Energy Mat Sol Cells 94:699–708
23. Berson S, De Bettignies R, Bailly S, Guillerez S, Jousselme B (2007) Adv Funct Mater 17:3363–3370
24. O'Regan B, Grätzel M (1991) Nature 353:737–740
25. Green MA, Emery K, Hishikawa Y, Warta W (2010) Prog Photovolt Res Appl 18:144–150
26. Ito S, Chen P, Comte P, Nazeeruddin KM, Liska P, Péchy P, Grätzel M (2007) Prog Photovolt Res Appl 15:603–612
27. Bai Y, Cao Y, Zhang J, Wang M, Li R, Wang P, Zakeeruddin SM, Grätzel M (2008) Nat Mater 7:626–630
28. Chiba Y, Islam A, Watanabe Y, Komiya R, Koide N, Han L (2006) Jpn J Appl Phys 45:L638–L640
29. Mishra A, Fischer MKR, Bäuerle P (2009) Angew Chem 121:2510–2536
30. Nazeeruddin MK, De Angelis F, Fantacci S, Selloni A, Viscardi G, Liska P, Ito S, Takeru B, Grätzel M (2005) J Am Chem Soc 127:16835–16847
31. Kuciauskas D, Monat JE, Villahermosa R, Gray HB, Lewis NS, McCusker JK (2002) J Phys Chem B 106:9347–9358
32. Baldo MA, O'Brien DF, You Y, Shoustikov A, Sibley S, Thompson ME, Forrest SR (1998) Nature 395:151–154
33. Sony Global - Sony Design - Activity - Sustainable Design, Dye-sensitized solar cell, 2013 http://www.sony.net/Fun/design/activity/sustainable/dssc.html
34. Eco-Products 2008: Sony solar-powered LED lamps, 2008 http://www.digitimes.com/photogallery/showphoto.asp?ID=3477&tag=25
35. Koops SE, O'Regan BC, Barnes PRF, Durrant JR (2009) J Am Chem Soc 131:4808–4818
36. Grätzel M, Durrant JR (2008) Nanostructured and Photoelectrochemical Systems for Solar Photon Conversion, Imperial College Press.
37. Ito S, Zakeeruddin SM, Humphry-Baker R, Liska P, Charvet R, Comte P, Nazeeruddin MK, Péchy P, Takata M, Miura H, Uchida S, Grätzel M (2006) Adv Mater 18:1202–1205
38. Gallo D, Landi C, Pasquino N (2008) Measurement 41:334–340
39. Kay A, Grätzel M (1993) J Phys Chem 97:6272–6277
40. Wang X-F, Xiang J, Wang P, Koyama Y, Yanagida S, Wada Y, Hamada K, Sasaki S-I, Tamiaki H (2005) Chem Phys Lett 408:409–414
41. Wang X-F, Kitao O, Zhou H, Tamiaki H, Sasaki S-I (2009) Chem Commun 1523–1525
42. Calogero G, Di Marco G, Caramori S, Cazzanti S, Argazzi R, Bignozzi CA (2009) Energy Environ Sci 2:1162–1172

43. Olea A, Ponce G, Sebastian PJ (1999) Sol Energy Mat Sol Cells 59:137–143
44. Calogero G, Di Marco G, Cazzanti S, Caramori S, Argazzi R, Di Carlo A, Bignozzi CA (2010) Int J Mol Sci 11:254–267
45. Campbell WM, Jolley KW, Wagner P, Wagner K, Walsh PJ, Gordon KC, Schmidt-Mende L, Nazeeruddin MK, Wang Q, Grätzel M, Officer DL (2007) J Phys Chem C 111:11760–11762
46. Yella A, Lee H-W, Tsao HN, Yi C, Chandiran AK, Nazeeruddin MDK, Diau EW-G, Yeh C-Y, Zakeeruddin SM, Grätzel M (2011) Science 334:629–634
47. Mozer AJ, Wagner P, Officer DL, Wallace GG, Campbell WM, Miyashita M, Sunaharac, K; Mori S (2008) Chem Commun 4741–4743.
48. Nüesch F, Moser JE, Shklover V, Grätzel M (1996) J Am Chem Soc 118:5420–5431
49. Martínez-Díaz MV, Ince M, Torres T (2011) Monatsh Chem 142:699–707
50. Yum J-H, Hagberg DP, Moon S-J, Karlsson MK, Marinado T, Sun L, Hagfeldt A, Nazeeruddin MK, Grätzel M (2009) Angew Chem 121:1604–1608
51. Karthikeyan CS, Thelakkat M (2008) Inorg Chim Acta 361:635–655
52. Yum J-H, Chen P, Grätzel M, Nazeeruddin MK (2008) ChemSusChem 1:699–707
53. Fabregat-Santiago F, Bisquert J, Cevey L, Chen P, Wang M, Zakeeruddin SM, Grätzel M (2009) J Am Chem Soc 131:558–562
54. Kron G, Egerter T, Werner JH, Rau U (2003) J Phys Chem B 107:3556–3564
55. Cahen D, Hodes G, Grätzel M, Guillemoles JF, Riess I (2000) J Phys Chem B 104:2053–2059
56. Han H, Bach U, Cheng Y-B, Caruso RA (2007) Appl Phys Lett 90:213510_3
57. Schwanitz K, Mankel E, Hunger R, Mayer T, Jaegermann W (2007) Chimia 61:796–800
58. Andersson A, Johansson N, Bröms P, Yu N, Lupo D, Salaneck WR (1998) Adv Mater 10:859–863
59. Park N-G, van de Lagemaat J, Frank AJ (2000) J Phys Chem B 104:8989–8994
60. Ito S, Murakami TN, Comte P, Liska P, Grätzel C, Nazeeruddin MK, Grätzel M (2008) Thin Solid Films 516:4613–4619
61. Peng B, Jungmann G, Jäger C, Haarer D, Schmidt H-W, Thelakkat M (2004) Coord Chem Rev 248:1479–1489
62. Ito S, Liska P, Comte P, Charvet R, Péchy P, Bach U, Schmidt-Mende L, Zakeeruddin SM, Kay A, Grätzel M, Nazeeruddin MK (2005) Chem Commun 34:4351–4353
63. Peter LM (2007) J Phys Chem C 111:6601–6612
64. Burke A, Ito S, Snaith H, Bach U, Kwiatkowski J, Grätzel M (2008) Nano Lett 8:977–981
65. Kavan L, Grätzel M (1995) Electrochim Acta 40:643–652
66. Snaith HJ, Grätzel M (2006) Adv Mater 18:1910–1914
67. Greene LE, Law M, Yuhas BD, Yang P (2007) J Phys Chem C 111:18451–18456
68. Barnes PRF, Anderson AY, Koops SE, Durrant JR, O'Regan BC (2009) J Phys Chem C 113:1126–1136
69. O'Regan BC, Durrant JR, Sommeling PM, Bakker NJ (2007) J Phys Chem C 111:14001–14010
70. Barbé CJ, Arendse F, Comte P, Jirousek M, Lenzmann F, Shklover V, Grätzel M (1997) J Am Ceram Soc 80:3157–3171
71. Zhang Z, Ito S, O'Regan B, Kuang D, Zakeeruddin MS, Liska P, Charvet R, Comte P, Nazeeruddin KM, Péchy P, Humphry-Baker R, Koyanagi T, Mizuno T, Grätzel M (2007) Z Phys Chem 221:319–327
72. Michaelson HB (1977) J Appl Phys 48:4729–4733
73. Snaith HJ, Moule AJ, Klein C, Meerholz K, Friend RH, Grätzel M (2007) Nano Lett 7:3372–3376
74. Bach U, Lupo D, Comte P, Moser JE, Weissörtel F, Salbeck J, Spreitzer H, Grätzel M (1998) Nature 395:583–585
75. Ding I-K, Tétreault N, Brillet J, Hardin BE, Smith EH, Rosenthal SJ, Sauvage F, Grätzel M, McGehee MD (2009) Adv Funct Mater 19:2431–2436
76. Schmidt-Mende L, Zakeeruddin SM, Grätzel M (2005) Appl Phys Lett 86:013504_1-3

77. Kroeze JE, Hirata N, Schmidt-Mende L, Orizu C, Ogier SD, Carr K, Grätzel M, Durrant J (2006) Adv Funct Mater 16:1832–1838
78. Snaith HJ, Grätzel M (2006) Appl Phys Lett 89:262114–1
79. Wang M, Xu M, Shi D, Li R, Gao F, Zhang G, Yi Z, Humphry-Baker R, Wang P, Zakeeruddin SM, Grätzel M (2008) Adv Mater 20:4460–4463
80. Schmidt-Mende L, Grätzel M (2006) Thin Solid Films 500:296–301
81. Schmidt-Mende L, Bach U, Humphry-Baker R, Horiuchi T, Muira H, Ito S, Uchida S, Grätzel M (2005) Adv Mater 17:813–815
82. Krüger J, Plass R, Cevey L, Piccirelli M, Grätzel M (2001) Appl Phys Lett 79:2085–2087
83. Howie WH, Harris JE, Jennings JR, Peter LM (2007) Sol Energy Mat Sol Cells 91:424–426
84. Würfel P (2009) Physics of solar cells, Wiley-VCH, Weinheim
85. An H, Xue B, Li D, Li H, Meng Q, Guo L, Chen L (2006) Electrochem Commun 8:170–172

Chapter 6
Construction of Solar Cells

6.1 Experimental Details

6.1.1 Characterization

The electrical characterization of the solar cells was done by measuring the I-V characteristic with the potentiostat *PGSTAT 10* from *Metrohm.* To compare the performance of the solar cellts under illumination an incandescent lamp with 60 W power and with a reflecting coating at the rear side (denoted as: the lamp), has been used in a standardized setup to ensure the same illumination conditions for all solar cells. The setup was located within a metal cage to exclude any influence of the environmental light. The solar cells were placed in a fixed central position 10 cm beneath the lamp, as shown in Fig. 6.1. Underneath the solar cell a sheet of black paper was placed to ensured that no light could be reflected back to pass the solar cell twice.

6.1.2 Materials and Sample Preparation

The solar cells were prepared on 1.1 mm thick borosilicate glass substrates coated with the fluorine doped tin oxide layer of approx. 550 nm thickness, the so-called FTO (SnO_2:F). These transparent conductive oxide layers were purchased from *Solaronix SA* and had a sheet resistance of about 10 Ω/square at a transmission of 80–90 % in the visible wavelength range. They were cleaned prior to use by sonication within an aqueous solution of the alkaline detergent *Deconex 24 forte* from *Borer* in a concentration of 0.74–1.0 g *Deconex* in 250 ml deionized water. After purging the samples with new deionized water, they have been sonicated within extra pure water (HPLC-grade from *Merck*) followed by sonication in ethanol (pro Anal., 96 %) from *Merck.* As final cleaning step the FTO substrates have been annealed at 500 °C for 1 h.

For DSSCs with a liquid electrolyte we used *Iodolyte AN-50,* purchased from *Solaronix.* The additive 4-*tert*-butylpyridine (tBP), which co-adsorbs onto the TiO_2

P. L. Marek, *Biomimetic Dye Aggregate Solar Cells*,
Springer Theses, DOI: 10.1007/978-3-319-00636-9_6,

Fig. 6.1 Photo of the I-V measurement setup. No light from outside could enter the dark cage with the incandescent lamp 10 cm above the solar cell, which was always placed in the same fixed lateral position. Marek PL, Hahn H, Balaban TS (2011) Energy Environ Sci 4:2366-2378 – Reproduced by permission of The Royal Society of Chemistry

surface, was used to increase the potential of the solar cells and to reduce the recombination rate at the TiO_2/electrolyte interface.

The solid hole transport material (HTM) we used for our ss-DSSCs was spiro-MeOTAD, purchased from *Merck*. It was dissolved within chlorobenzene to a concentration of 0.15–0.17 M with the addition of 0.12–0.13 M *tert*-butylpyridine (tBP) and 8.3–13 mM bis(trifluoromethane)sulfonimide lithium salt (TFSI-Li), according to the literature [1–3]. A small overview about common fabrication methods and additive concentrations is given in the appendix.

Beside the investigation, if P2 aggregates can compete with dye monolayers by the sensitization of TiO_2, two other technologies have been introduced and investigated for the solar cell fabrication. One was the atomic layer deposition (ALD) as application method for the dense hole blocking TiO_2 layer, instead of the common spray pyrolysis deposition (SPD) technique. The other method was the application of the gold counter electrode by sputtering, instead of the conventional evaporation technique.

ALD is a vacuum-based process, similar to chemical vapor deposition (CVD) with the difference, that the grown layer thickness can be controlled on the atomic scale by pulsing two gaseous reactants, typically a metalorganic precursor and water, alternatively over the surface of a heated substrate. After chemisorption of a monolayer of the metal precursor no further growth process takes place until the gaseous water is let in and reacts with this metalorganic monolayer forming a metaloxide monolayer. Onto this surface, the next metalorganic monolayer can be chemisorbed and so on. This self-limiting growth process results in highly uniform and dense coatings with a thickness precisely controlled by the number of cycles.

For the first TiO_2 layers, we used titanium tetrachloride ($TiCl_4$) as precursor, which grows at a rate of 0.05 nm per cycle with water as second reactant at a substrate temperature of 250 °C [4]. Later on the precursor was changed to titanium isopropoxide (Ti-iProp.) because the new precursor was better suited for our ALD system in concerns of contamination and corrosion, which was caused by HCl, a byproduct of the $TiCl_4$ decomposition. According to the literature, the TiO_2 layers grown from Ti-iPorp. were of a similar quality as those from the $TiCl_4$ precursor [5]. The typical growth rate for Ti-iProp. is 0.03 nm/cycle [6]. To ensure the maximum degree of crystallization the growth temperature has been increased to 300 °C, which is the highest growth temperature for the self-limiting growth process with Ti-iProp. The oxygen vacancy concentration is known to increase with decreasing temperature up to a few atom- % at temperatures as low as 100 °C, leading to amorphous $TiO_{2\text{-}x}$ layers [7].

The DC magnetron sputter coater *108 auto* from *Cressington* was used to sputter 50 nm thick layers of pure gold at 20 mA. The layer thickness was monitored with a quartz crystal oscillator. The samples were covered by a mask and placed not closer than 4 cm beneath the gold target to avoid heating.

6.2 Solar Cell Preparation and Investigation

6.2.1 Dye (Aggregate) Sensitized Solar Cells with a Liquid Electrolyte

P2 was first implemented into solar cells using the architecture of conventional dye sensitized solar cells (DSSCs) with a liquid electrolyte. They do not require inert gas fabrication techniques and are not very sensitive for small defects and inhomogeneities in the layer structure, because the Fermi level of the FTO gets pinned to the redox potential of the iodide couple (I^-/I_3^-) of the electrolyte. A dense TiO_2 underlayer as hole blocking layer on the FTO is therefore not essential, but helps to improve the performance of DSSCs [8]. They have been used as first model structures to test P2 aggregates as sensitizers for mesoporous nanocrystalline TiO_2 and the effect of the dense TiO_2 underlayer.

Some FTO substrates were coated with a 40 nm thick dense TiO_2 underlayer (UL) by ALD using the precursor titanium tetrachloride ($TiCl_4$). A mesoporous

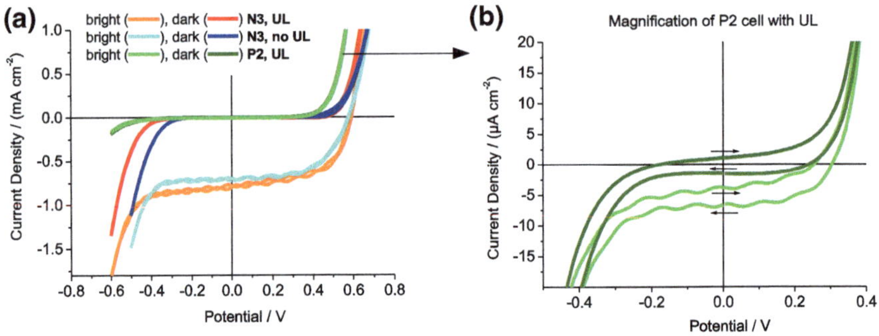

Fig. 6.2 I-V curves of DSSCs with P2 and N3. **a** I-V curves in the dark (*dark colors*) and under illumination with the lamp (bright colors) of two DSSCs with the ruthenium dye N3 sensitizing a mesoporous TiO_2 layer with (*reddish*) and without (*blueish*) a dense TiO_2 underlayer (UL). The *green* traces show the I-V curves of a DSSC sensitized with P2 (with UL). They are magnified in (**b**) with indicated scan directions. The sweep in both directions caused the hysteresis at a scan rate of 0.25 V/s, the wiggles originated by a too small step potential (5 mV/step), thus too short integration time for each point. P2 was sprayed from THF, followed by a post-treatment within *n*-heptane. The mesoporous TiO_2 layer contained additional 1-µm sized macropores. Short summary: FTO/[ALD-TiO_2 (UL)]/mesoporous TiO_2 + 1 µm pores/dye [N3 or P2] / I^-/I_3^- electrolyte/Pt-FTO

TiO_2 layer has been doctor-bladed onto these substrates. The aqueous dispersion was prepared by mixing TiO_2 nanoparticles from *Evonik* (P25) with water and with an aqueous dispersion of 1-µm-sized polystyrene spheres in order to obtain additional 1-µm-sized macropores within the mesoporous layer. The macropores, which made up about 10 vol. % of the TiO_2, were used for investigations of the pore filling ratio and morphology of P2 within the inner layer [9]. They had the additional advantage of preventing the film from cracking during the sintering process, what is normally achieved with other additives like ethyl cellulose. Furthermore, such macropores are known to enhance the efficiency due to acting as light scattering centers [10]. During sintering at 450 °C for 30 min, the polystyrene spheres were pyrolyzed leaving behind a mesoporous TiO_2 layer with embedded macropores. As reference the same structures have been stained by the common ruthenium dye N3 (purchased from *Solaronix*). P2 has been applied onto the porous layer by spraying from a 0.4 mM P2 solution in THF. Afterward, the P2 samples have been annealed for 2.5 h within *n*-heptane at its boiling temperature of 98 °C. As counter electrode served a second FTO substrate, which was coated by a thin and highly transparent Pt layer as catalyst using a doctor-blading paste from *Solaronix*. Both electrodes, the stained one and the counter electrode were sealed together with a 60 µm thick hot melting thermoplast foil from *Solaronix*. The gap was filled with the liquid electrolyte (*Iodolyte AN-50* from *Solaronix*) through holes which were drilled in advance. The corresponding I-V curves in the dark and under illumination with the lamp are shown in Fig. 6.2.

It can be seen that the performance of the DSSCs with the N3 dye did not change significantly with the TiO_2 UL. Only the breakdown voltage in the third quadrant shifted slightly to more negative potentials with the UL. The efficiency of the DSSC with the P2 aggregates was considerably lower. And also the open circuit voltage was lower by about 0.1 V. The forward and backward sweep of the applied potential, indicated by the arrows (b), caused a hysteresis at these low currents.

The wiggles in the measurement under the illumination are an artifact caused by short settling times between the measurement steps due to a fast scan rate at small step potentials, which were here 0.25 V/s and 5 mV/step, respectively. These wiggles could be eliminated for the following measurements by stetting the step potential to 20 mV.

The main reason for the low efficiency of the P2 sensitized solar cells was assumed to be a dissolution of P2 within the electrolyte.

6.2.2 All-Solid-State Solar Cells

The first all-solid-state solar cells with P2 have been prepared on bare flat TiO_2 layers of 40 nm thickness, grown by ALD onto a partly etched FTO layer. 1.8 ml of a 0.1 mM dispersion of P2 in anhydrous DCM was sprayed onto this FTO-TiO_2 substrate. After spin-coating a solution of the solid HTM spiro-MeOTAD (0.15 M spiro-MeOTAD, in chlorobenzene containing also 9.3 mM TFSI-Li and 0.12 M tBP) a 100 nm thick gold electrode has been evaporated on top forming an active solar cell area of 0.69 cm^2. The solar cell has been characterized by I-V measurements under the illumination with a solar simulator, which was located in a N_2-glovebox (Fig. 6.3).

The efficiency was 0.01 % on an active area of 0.69 cm^2. Other parameters were:

V_{OC}: 0.46 V, I_{SC}: –0.045 mA/cm^2, V_{MPP}: 0.3 V, I_{MPP}: 0.031 mA/cm^2, FF: 44.3 %

Although the efficiency was very low, it was an improvement of over one order of magnitude compared to the DSSC with the liquid electrolyte. The open circuit voltage did also significantly increase.

As such solar cells were very sensitive to defects in the surface, either intrinsic or induced by handling the mask for the evaporation, it was thought about a new cell design with miniaturized solar cells.

6.3 New Solar Cell Design

To reduce the risk that the solar cells get short-circuited at local inhomogeneities or defects and to improve the statistics for a higher validity of the measurements, the solar cells have been miniaturized on the substrate. Therefore, a mask has been used to apply 16 identical gold counter electrodes, each forming one solar cell of 7×1 mm^2 area on a 16×16 mm^2 substrate. This structure allowed also the

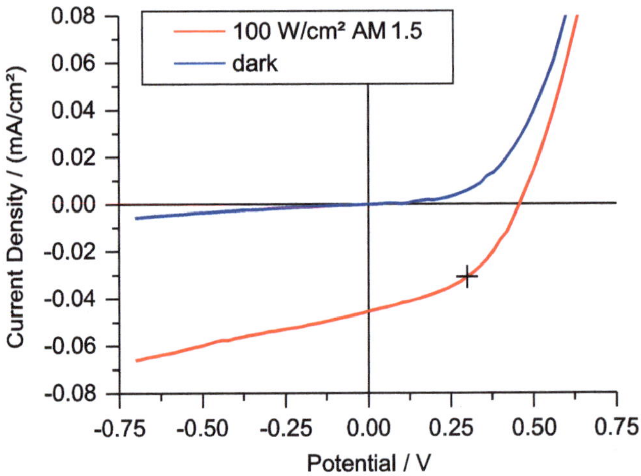

Fig. 6.3 I-V curve of a flat solar cell with P2 sprayed from DCM. I-V characteristic of a probably amorphous P2 film, which was sprayed from DCM onto a flat TiO_2 layer grown by ALD. It was measured under simulated sunlight (100 mW/cm^2 AM 1.5). The cross marks the maximum power point (MPP). Short summary: FTO/ALD-TiO_2 (40 nm)/P2 sprayed from DCM/ spin-coated spiro-MeOTAD/100 nm Au evaporated

application and characterization of different solar cell types on one substrate, for example, when only one half of the substrate was covered an additional porous layer.

This solar cell array on the substrate required a precise contacting method for the reproducible and fast contacting or switching between the electrodes. Therefore, a contacting device has been developed, which contacted all solar cells on one substrate at once by 18 golden spring probes, which maintained the same contact pressure for each cell. Two rotary switches were used to switch between the contacting wires, which were all of the same length to keep the series resistance constant. The rotary switch itself had a contacting resistance below 50 mΩ, what was negligible compared to the series resistance of the solar cells. Additionally, it has been found, that the contact over the spring probes was soft enough to allow the reproducible contacting without the necessity to etch away the FTO layer underneath the gold electrode area as this was not short-circuited by contacting the gold electrode on top. This allowed to keep the active cell area precisely constant to 7 mm^2, as it was predefined only by the gold sputtering mask. Two spring probes were free to contact the FTO substrate. Therefore, the surface was scratched within the contact area with a scalpel and painted with a drop of silver paint, a dispersion of silver particles. The contact of the gold spring probe with the dried silver paint ensured a reproducible ohmic contact to the FTO electrode. It made no significant difference at which of both opposing sides the FTO was contacted, thus only one of both redundant spring probes was used as counter contact for all 16 solar cells on the substrate. After each contacting the

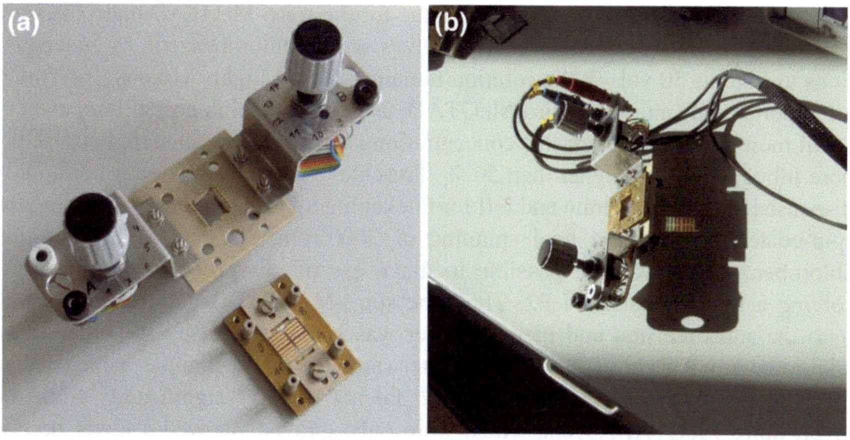

Fig. 6.4 New solar cell design and the contacting device. **a** The substrate with the 16 active solar cell areas, mounted with the active coating on top is fixed onto the holder by two screws (bottom right). It is being put up side down onto the contacting device (top left) guided by four rods and fixed by two diagonally sitting screws. The contact is established by the gold spring probes of the contacting device. **b** The solar cells are illuminated through the hole in the substrate holder, ready contacted for a test-measurement in the sunlight spot, which is transmitted through the semitransparent substrate. Marek PL, Hahn H, Balaban TS (2011) Energy Environ Sci 4:2366-2378 – Reproduced by permission of The Royal Society of Chemistry

spring probes were wiped clean with a paper towel soaked in ethanol, to ensure a contact reproducibility approaching 100 %.

Fig. 6.4 shows the contacting device and a reference solar cell based on the common ruthenium dye N719.

Each of both rotary switches addressed eight solar cells. The shown solar cell substrate was partly coated with a mesoporous TiO_2 layer, which got markedly stained by N719 (brownish color). A staining of the remaining dense TiO_2 underlayer could not be noticed by the naked eye, as can be seen in the button right part of the solar cells in (a) and as pale part of the sunlight spot transmitted through the solar cells in (b). Dye aggregates instead could also markedly stain flat TiO_2 layers.

6.3.1 Solar Cells with Amorphous P2 Infiltrated into Mesoporous TiO_2

A P2 sensitized solar cell was prepared on the mesoporous TiO_2 (T37) layer, which had a 30 nm thick TiO_2 underlayer, applied by ALD. From now on titanium isopropoxide (Ti-iProp.) was used to prepare the dense TiO_2 underlayers on the FTO substrates by ALD, replacing the previously used $TiCl_4$. 0.4 ml of a 0.2 mM solution of P2 in THF was sprayed onto the substrate.

The most common application technique for the spiro-MeOTAD matrix is spin-coating. Although spiro-MeOTAD dissolves within chlorobenzene to concentrations as high as 50 vol. % the solution remains still not highly viscous [2]. This is an important advantage of spiro-MeOTAD, as it allows a high pore filling, even of small mesopores. However, this concentration is still not high enough, because the pore filling should be higher than 50 %. To achieve this, the solution is commonly dispensed onto the substrate and left for thickening up to 1 min prior to starting the spin-coater, mostly to a final spinning of 2000 rpm. As P2 is soluble within chlorobenzene, it was not possible to make use of this procedure without dissolving a major amount of P2. Thus, the spiro-MeOTAD solution was sprayed onto the substrate first and then the layer was smoothed out by spin-coating an additional small drop of the solution afterward. By this technique, which led to macroscopically homogeneous films, P2 did not dissolve significantly from the layer. Spiro-MeOTAD layers, which were only sprayed, showed a lower performance, probably because they did not cover the TiO_2 surface completely. About 30 μl of the spiro-MeOTAD solution were sprayed on top, followed by the spin-coating of additional 20 μl of the same solution at 2000 rpm for 25 s. The spiro-MeOTAD solution contained 0.15 M spiro-MeOTAD in chlorobenzene and 8.3-mM lithium salt (pre-dissolved in acetonitrile to 170 mg/ml) and 0.12 M *tert*-butylpyridine as additives.

Afterward a 50 nm thick gold layer was sputtered through a mask, forming 7 mm^2 large stripes. This application technique for the gold layers replaced from now on the previously used evaporation technique. As the FTO layer underneath the gold contact area had been etched away before applying the dense TiO_2 underlayer, the active area (intersection of both electrode areas) was 5.5 mm^2 for each cell.

One representative I-V characteristic, measured under illumination with the lamp, is shown in Fig. 6.5.

Considering the lower irradiation intensity of the lamp, compared to that of sunlight, the present I-V curve revealed a significant increase in I_{SC} (0.054 mA/cm^2) compared to that of the P2 sensitized flat TiO_2 solar cell. Also the V_{OC} slightly increased up to 0.50 V.

The solar cell performance did not decrease significantly with the irradiation time, as can be seen in the plot of the short circuit current over time, where the lamp has been switched on and off three times (b). The shunt resistance of the illuminated cell was 38 kΩ cm^2, what is relatively low, hence indicating a relative high recombination rate.

For comparison a reference cell with the most common ruthenium dye N719 (purchased from *Solaronix*) was prepared on an identical substrate, as described in the next paragraph.

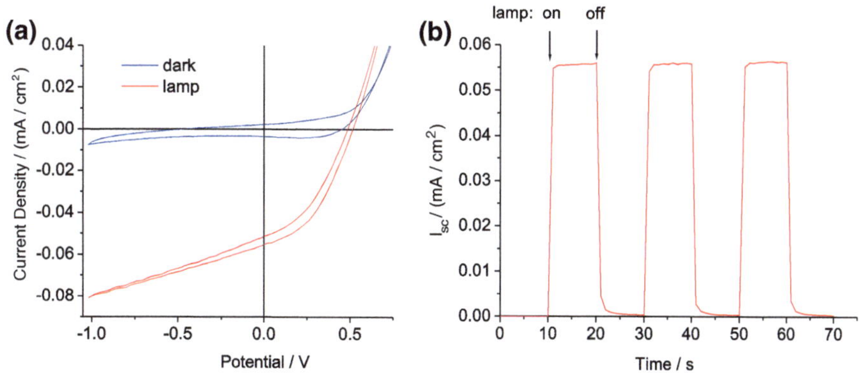

Fig. 6.5 I-V curves and time plot of a mesoporous ss-DSSCs with P2 sprayed from THF. **a** I-V curves in the dark and at illumination by the lamp, measured at a scan rate of 0.2 V/s. **b** Corresponding short circuit current (I_{SC}) over time during three cycles of switching the lamp on and off. P2 was sprayed from a solution in THF onto a mesoporous TiO_2 layer on a dense TiO_2 underlayer. The spiro-MeOTAD solution was first sprayed and afterward spin-coated on top. A total of 50 nm thick gold electrodes were sputtered to 7 mm^2 large stripes. As the FTO underneath was partly etched away the active areas were 5.5 mm^2 large. Short summary: FTO/ALD-TiO_2 (30 nm)/mesoporous TiO_2 (T37)/2 h at 450 °C/P2-THF sprayed/spiro-MeOTAD sprayed/spiro-MeOTAD spin-coated/50 nm Au sputtered

6.3.2 1st ss-DSSC Reference Solar Cell with the Dye N719

An identical mesoporous (T37) substrate as used for the previous P2 solar cells was sensitized by the reference dye N719. The substrate was dried at 110 °C for 2 days and put hot into the N719 staining solution, which contained 0.5 mM N719 in a mixture of *tert*-butanol and acetonitrile (1:1 volume ratio). After 8.5 h of staining at room temperature, the sample has been taken out and purged over night within pure acetonitrile to desorb dye agglomerates, which are known to decrease the efficiency. The same spiro-MeOTAD solution as used for the P2 ss-DSSCs has been spin-coated according to the literature [2]: 20 μl of the MeOTAD solution have been dispensed on the surface and after 40 s of thickening the spin-coater was started to run at 2000 rpm for 25 s. At the end, the 50 nm thick gold electrode was sputtered forming 7 mm^2 large stripes. They were defining the whole active area, because the FTO layer underneath was not etched away in advance. It has been figured out, that this was not necessary because the contacting device did not short circuit the cells by contacting the top gold electrode in the area with the FTO layer underneath (separated by the spiro-MeOTAD/P2/TiO_2 layer).

Beside the I-V measurement at the illumination with the lamp an I-V curve has also been measured under the illumination with direct sunlight through an open window. Representative I-V curves of the 1st N719 reference ss-DSSCs are given in Fig. 6.6.

A kink can be seen in the fourth quadrant of the I-V curve, especially under the illumination by the lamp (red). The shunt resistance decreased significantly at the

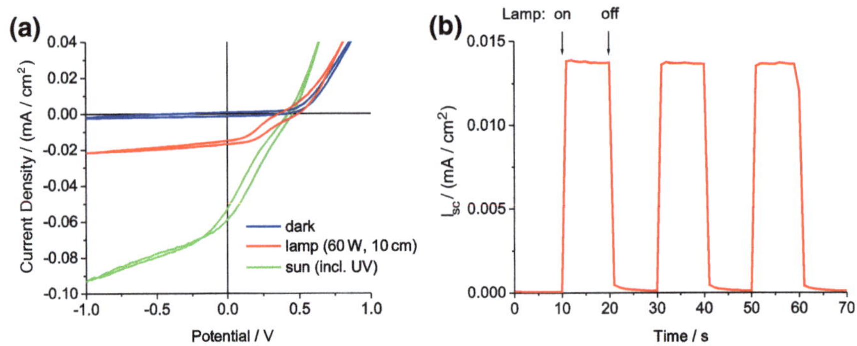

Fig. 6.6 I-V curves and time plot of the 1st N719 reference ss-DSSC. **a** I-V curves of the 1st N719 reference solar cell on an identical substrate as for the P2 cell from Fig. 6.5, under the illumination with the lamp (*red*) and in direct sunlight (*green*). Scan rate: 0.2 V/s. **b** I_{SC} three times irradiated with the lamp. Short summary (differences highlighted): FTO/ALD-TiO_2 (30 nm)/mesoporous TiO_2 (T37)/2 h at 450 °C/N719/spiro-MeOTAD spin-coated/50 nm Au sputtered

higher irradiation intensity of the sunlight (green), reaching 35 kΩ cm^2, whereas that under the illumination with the lamp was with 175 kΩ cm^2 five times higher. The time-plot revealed that the short circuit current slightly decreased upon the exposure to the light of the lamp. The overall photocurrent was quite small, what may be due to the long (over night) purging time within acetonitrile. This may have led to a partial desorption of not only dye agglomerates, resulting in a weak staining.

Snaith et al. explained the occurrence of similar kinks with a high n-doping concentration of the TiO_2 underlayer, which is caused by oxygen vacancies [11]. When the elevated Fermi level aligns with the lower lying one of the FTO (with respect to the vacuum level) a depletion region occurs in the TiO_2 forming a Schottky barrier at the interface with the FTO, which can be considered as metallic. A similar band model for the interface between TiO_2 and FTO in DSSCs, showing also the case of a very low-doped TiO_2, was described by Kron et al. [12]. Rühle et al. developed a theoretical model for the charge transport through such an FTO-TiO_2 Schottky barrier [13]. It is dominated by tunneling, due to the narrow band banding region, originating by a high doping concentration of the TiO_2. Snaith also reports that this barrier may be reduced by the irradiation with UV light, what is also called the UV effect of such solar cells. The UV light is supposed to form surface states within the gap of the TiO_2 which pins the Fermi level of the FTO to this lower level. Snaith observed a strong kink upon the illumination of conventional ss-DSSCs with monochromatic light, thus without a UV content. We might face the same effect with the 60 W incandescent lamp, which has also a very low fraction of UV light.

Therefore, we tested these solar cells also for the UV effect by the illumination with natural direct sunlight, passing through an open window in order to not significantly cut off the UV part. The effect was, however, quite small and did not lead to a significant decrease of the kink and increase of the fill factor. The edge in the I-V curve, at which the photocurrent saturates was strongly shifted toward negative potentials, even into the third quadrant.

Snaith et al. demonstrated also that the use of pure oxygen as carrier gas for the application of the TiO_2 underlayer by the common SPD technique significantly reduced the kink encountered upon monochromatic illumination. However, even when the TiO_2 layers were prepared under pure oxygen (at ambient pressure), the I-V curves still showed a slight kink, when the solar cells were illuminated by monochromatic light. Whereas under simulated sunlight no kink was present and the fill factor could be significantly increased up to 0.73 when the TiO_2 underlayer was prepared under pure oxygen, a kink arose also under simulated sunlight when the TiO_2 underlayer was prepared by SPD with argon as carrier gas. [11] The kink in our cells has probably the same origin, but it is much more pronounced and not even UV light is capable to reduce it significantly. As the doping level of TiO_2 is known to rise already upon the exposure to vacuum or inert gas [14] and our ALD process is run under vacuum with nitrogen as carrier gas, we expect a very high oxygen vacancy concentration in the as prepared TiO_2 underlayers. When the oxygen depletion is supposed to increase upon the exposure to a low oxygen partial pressure it might be also possible to replenish the TiO_{2-x} with oxygen by post-annealing within an oxygen rich atmosphere. Therefore, the samples were annealed longer within air. Additionally, a surface treatment of the TiO_2 layer was performed within an aqueous solution of $TiCl_4$. This method leads to the growth of a 1–2 nm thick TiO_2 overlayer on the whole substrate [15]. It is commonly applied to improve the TiO_2 surface of mesoporous TiO_2 layers, which leads to an improvement of the solar cell performance. The effect of these treatments on the I-V characteristic was tested in a 2nd N719 reference ss-DSSC.

6.3.3 2nd N719 Reference ss-DSSC with Intensified Thermal Treatment

To facilitate the oxygen diffusion during post-annealing in air, the thickness of the TiO_2 underlayer was reduced to 23 nm by running 700 cycles with Ti-iProp. instead of 850. An additional annealing step at 450 °C for 2 h in air was added directly after the growth of the TiO_2 underlayer. The mesoporous layer (T37) was doctor-bladed afterward and sintered at 500 °C for 1 h. A subsequent treatment of the substrate was performed within a 0.02 M solution of $TiCl_4$ in distilled water for 7 h at room temperature to further improve the TiO_2 surface, as this treatment is supposed to grow an additional 1–2 nm thick TiO_2 overlayer. After this treatment, a third annealing step was performed also at 500 °C for 1 h in air.

This 2nd N719 reference solar cell was stained within the same N719 solution, which has been used also for the 1st N719 solar cell, at room temperature in the dark. Afterward, it has been rinsed and kept for approx. 30 min within the pure solvent mixture of *tert*-butanol and acetonitrile (1:1 volume ratio) in order to gently desorb possible dye agglomerates. The application of the spiro-MeOTAD matrix and sputtering of the 7 mm^2 large gold electrodes was kept the same as for the first cell. The TFO layer underneath the gold contact area was also not etched away. Thus, the active solar cell area was exactly defined by the gold area. The I-V characteristic of the 2nd N719 reference cell is shown in Fig. 6.7.

These I-V curves, under the illumination with the lamp (red) and in direct sunlight (green) still reveal a pronounced kink. However, the photocurrent increased significantly by a factor of 8 and also V_{oc} increased. A slightly faster degradation was observed in the measurement of the short circuit current over time.

The increase in photocurrent can be attributed to a higher dye amount adsorbed onto the TiO_2 surface, which depends also on the surface roughness. The treatment in $TiCl_4$ is assumed to increase the dye uptake and to increase the electron injection rate [16]. However, the kink, which is the most important factor decreasing the fill-factor and efficiency, remained almost unaffected.

The electrical parameters of the N719 reference solar cells are summarized in the next chapter in Table 6.1.

Glatthaar et al. observed also a strong kinking in the I-V curve of organic bulk heterojunction solar cells, which contained TiO_2 as hole blocking layer, which has been applied via electron beam evaporation onto a titanium electrode [17]. From impedance spectroscopy measurements these authors could exclude the effect of a counter diode and proposed as reason for this kink a hindering of the charge injection at one of the electrodes. This agrees with the model from Rühle and Cahen and Snaith et al., who explained the hindering of the charge injection by the

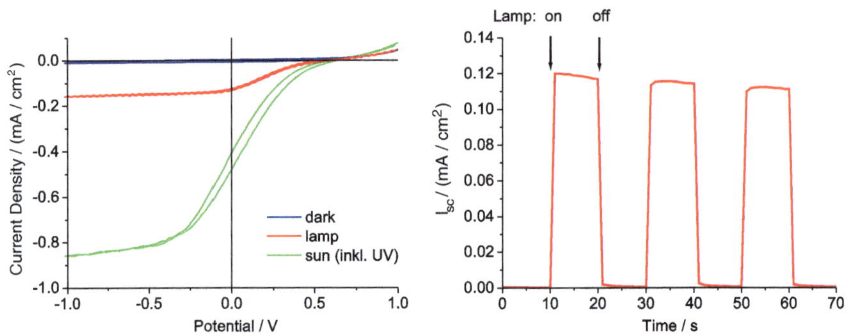

Fig. 6.7 I-V curves and time plot of the 2nd N719 reference cell. This 2nd N719 reference ss-DSSC differed from the first one in Fig. 6.6 by a thinner TiO_2 underlayer, intensified annealing in air and an additional treatment of the TiO_2 surface within an aqueous $TiCl_4$ solution, as can be seen highlighted in the summary below. The illumination by direct sunlight (green) was done through the open window at clear sky. Scan rate: 2 V/s. FTO/ALD-TiO_2 (23 nm)/2 h 450 °C/ mesoporous TiO_2 (T37)/1 h 500 °C/$TiCl_4$/1 h 500 °C/N719/spiro-MeOTAD spin-coated/50 nm Au sputtered

Table 6.1 Electrical parameters of the N719-based ss-DSSCs

ss-DSSC	Age	Illumi-nation	P (μW/cm^2)	FF (%)	I (−1 V) (μA/cm^2)	I_{sc} (μA/cm^2)	V_{oc} (V)	R_s (+1 V) (kΩ·cm^2)	R_{sh} (kΩ·cm^2)	$R_{s,\ center}$ (kΩ·cm^2)
1st N719	1 week	Dark	–	–	–	–	–	6.7	583	–
1st N719	1 week	Lamp	2.2	31	22	16	0.44	5.3	175	16.7
1st N719	1 week	Sun	5.8	24	93	56	0.44	3.0	35	6.6
2nd N719	2 weeks	Dark	–	–	–	–	–	5.8	175	–
2nd N719	2 weeks	Lamp	14.7	20	159	127	0.58	4.2	54	3.3
2nd N719	2 weeks	Sun	41.6	15	858	444	0.62	3.0	12	0.8
2nd N719	5 months	Dark	–	–	–	–	–	9.7	438	–
2nd N719	5 months	Lamp	2.9	15	107	40	0.49	7.8	70	5.3
2nd N719	5 months	SS	7.0	11	884	112	0.55	10.0	12	0.8
3rd N719	7 months	Dark	–	–	–	–	–	77.8	1400	–
3rd N719	7 months	Lamp	5.2	13	166	73	0.55	58.3	50	3.3
3rd N719	7 months	SS	7.8	12	1433	129	0.51	41.2	5	0.6

Representative values of the 1st, 2nd, and 3rd N719 solar cells. The parameters of the 2nd N719 cell are compared in the stage when it was 2 weeks and 5 months old. The current (I) was also measured at −1 V, because the kink distorted the explanatory power of I_{sc}. The series resistance (R_s) was measured always at +1 V. $R_{s,\ center}$ assigns the series resistance in the center of the I-V curve, that is, in the linear regime between quadrant III and IV. All solar cells had an area of 7 mm^2 . SS: solar simulator (100 mW/cm^2 AM 1.5) formation of a Schottky barrier at the interface with the FTO. The similar observation from Glatthaar et al. in the system comprising an interface between Ti and TiO_2, which is most probably n-doped, too, due to the vacuum-based application method, validates this model.

We observed further, that this kinking extends after a prolonged storage in the dark in air, as can be seen on the same cell after 5 months in Fig. 6.8.

The 2nd N719 cell has been measured 2 weeks after preparation under the lamp and in natural sunlight and again 5 months later under the lamp and simulated sunlight. After this storage in the dark and air the photocurrent decreased significantly and the kink extended toward higher negative potentials. This went along with an increase of the series resistance, which can be seen as smaller slope in quadrant I, but also as smaller slope in the center of the I-V curve between quadrant III and IV when comparing both measurements under illumination with the lamp (a). The illumination intensity of the solar simulator (b, dashed green trace) was slightly higher than that of the natural sunlight (solid green traces), which fell on the cell through the open window at clear sky in the afternoon. The fact, that both I-V curves reach the same current density, one at lower natural illumination intensity before and one at higher intensity after degradation of the cell, shows, that the extended kink is not related to a lower current density, but probably to the increased series resistance. Before degradation of the cell, which occurred partly during the measurements, but more significant during storage, the kink was minimal pronounced at low current densities as can be seen at the I-V curve of the 1st N719 solar cell (Fig. 6.6). The solar simulator used here was the model *81150* from *LOT-Oriel* with the sourcemeter *2400* from *Keithley*.

Normally, ruthenium dyes like the N719 are known to be very stable in DSSCs. The spiro-MeOTAD in ss-DSSCs has also proved a high stability in tests at 1000 h under illumination with visible light at 60 °C [18].

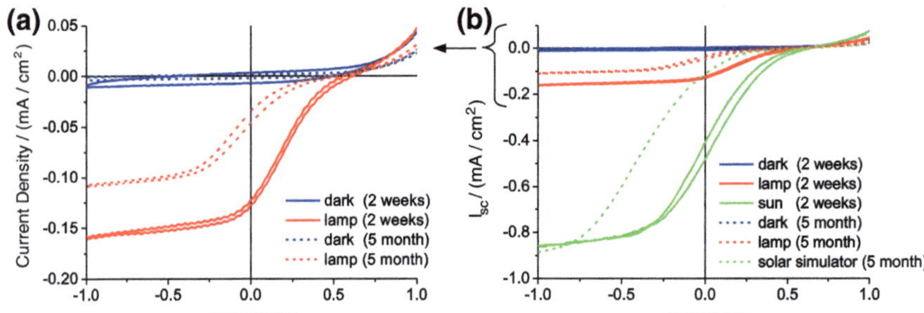

Fig. 6.8 I-V curves of the 2nd N719 cell after 5 months and in a solar simulator. The 2nd N719 reference from Fig. 6.7 was measured 2 weeks after preparation (solid traces) and after 5 months storage in the dark (*dashed traces*). **a** Magnification of the I-V curves measured in the dark (*blue*) and under illumination with the lamp (*red*). **b** The same cell has been measured also in natural sunlight after 2 weeks (*green* solid trace) and in a solar simulator after 5 months (*green dashed trace*, 100 mW/cm^2 AM 1.5)

As the post-annealing in air had no significant effect on the kink, we investigated the effect of a post-annealing within pure oxygen at ambient pressure. Instead of the surface treatment by $TiCl_4$·aq, we investigated the effect of smaller TiO_2 particles, which should increase the dye-uptake in order to increase the current–density.

6.3.4 3rd N719 Reference ss-DSSCs with Smaller TiO_2 Particles

For the 3rd N719 ss-DSSC 13 nm sized TiO_2 particles (T13) were used instead of the T37 with 37 nm particles and the ALD grown TiO_2 underlayer was annealed at 500 °C in pure oxygen, flowing at atmospheric pressure, for 1 h. This resulted in a further increase of the photocurrent, even without the treatment in the aqueous $TiCl_4$.

6.3.4.1 Preparation Details

The staining of the mesoporous TiO_2 substrate was done in the same N719 solution at 65 °C for 2 h in order to enhance the diffusion into the smaller mesopores. Afterward, the samples were rinsed with the pure solvent mixture (*tert*-butanol/AN) and immersed into it for 1 h at room temperature. A slightly higher concentrated spiro-MeOTAD solution has been prepared with 0.17 M spiro-MeOTAD, 13 mM TFSI-Li, and 0.12 M tBP in chlorobenzene. The TFSI-Li was pre-dissolved within acetonitrile according to the literature (170 mg TFSI-Li in 1 ml AN) and added together with 2.2 vol % AN into the spiro-MeOTAD solution [2]. As it was more difficult on this fine-porous layer to dispense 40 μl of the spiro-MeOTAD solution over the sample prior to spin-coating, three drops have been cast directly onto the spinning sample at 2000 rpm, resulting in a macroscopically homogeneous coating. Afterward, a 50 nm thick gold layer was sputtered through a mask forming 7 mm^2 large solar cells (the FTO layer underneath the gold contact area was not etched away).

Fig. 6.9 shows a representative solar cell of the 3rd N719 reference cells. The photocurrent could be further increased compared to the 2nd N719 reference cell. However, despite the annealing at 500 °C in pure oxygen the kink is not significantly reduced and hence also the efficiency. A degradation with time could be also observed for the third cell after a storage for 7 months (in the dark). In the photo of the solar cell substrate (b) the brownish area corresponds to the N719-stained mesoporous TiO_2 layer. The remaining flat ALD-grown TiO_2 layer was not markedly stained (right hand side).

Although the very fine-porous TiO_2 layer had a poor wetting ability for the spiro-MeOTAD solution the highest efficient solar cells could be realized with it.

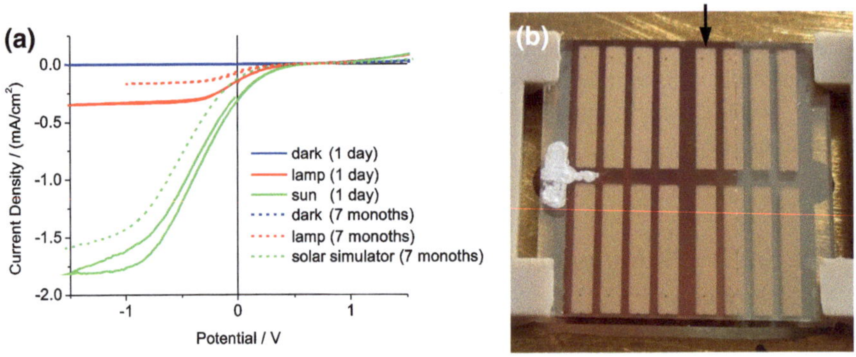

Fig. 6.9 I-V curves of a representative solar cell from the 3rd N719 substrate with photo. **a** I-V curves of a representative solar cell 1 day after preparation (solid traces) and after 7-month storage in the dark in air at room temperature (*dashed traces*). Shown are the measurements in the dark (*blue* traces), under the lamp (*red* traces), in natural sunlight (*green*, solid trace) and under a solar simulator (*green*, dashed trace, 100 mW/cm^2 AM 1.5) at scan rates between 0.5 and 1 V/s. **b** Photo of the substrate with the corresponding solar cell, which is marked by an arrow. Marek PL, Hahn H, Balaban TS (2011) Energy Environ Sci 4:2366-2378 – Reproduced by permission of The Royal Society of Chemistry FTO/ALD-TiO_2 (25 nm)/1 h 500 °C in O_2/mesoporous TiO_2 (T13)/2 h 500 °C/N719/spiro-MeOTAD spin-coated/50 nm Au sputtered

Spiro-MeOTAD is the HTM of choice for this challenge, as it combines a reasonable high conductivity with a relatively small molecular size. Additionally, it has a very high solubility while maintaining a low viscosity of the solution, even at concentrations around 50 vol % in chlorobenzene [2]. Such requirements for HTMs make it difficult to find alternatives, which might be necessary if the conductivity or the recombination rate could not be further improved by additives.

The following table compares the most important electrical parameters of all three N719 ss-DSSCs.

The series resistance did rise with the storage time, as shown for the 2nd N719 cell, leading to an extension of the kink toward negative and positive potentials. This reduced the fill-factor and with it significantly the power output. An almost linear regime of the series resistance was reached for fresh cells around 1 V, while the corresponding values of the cells after a few months aging were shifted toward higher positive potentials. A series resistance of about 2 kΩ cm^2 was reached at 1.5 V for the 5-month-old 2nd N719 cell and at 3 V for the 7-month-old 3rd N719 cell.

Due to the kink, the efficiency of these reference solar cells was very low, being 0.008 % for this 3rd N719 cell after 7-month storage. Without the kink instead, its high photocurrent at negative bias would let expect a theoretical efficiency almost two orders of magnitude higher as the efficiency scales with I_{SC} and FF.

$\eta = FF \cdot I_{SC} \cdot V_{OC}/(100\ \text{mW}) = 12\ \% \cdot 0.129\ \text{mA} \cdot 0.51\ \text{V} \approx 0.008\ \%$.

The current density was measured at −1 V, just in front of the steep decrease due to the kink, to be more than 11 times higher, than I_{SC} (at 0 V), hence I_{SC} is expected to have been reduced by over one order of magnitude due to the kink, and

FF, is expected to be decreased by the kink also several times. Additionally, in Fig. 6.9 can be seen, that the initial current density, measured under the lamp irradiation 1 day after cell preparation decreased by a factor of two during the 7-month-aging period. The measurement under natural sunlight was done for qualitative comparison reasons only and not for a quantitative analysis.

As the P2 dye aggregate solar cells (DASCs) seemed to be less affected by the formation of a kink in the I-V curve (Fig. 6.5) than the N719 reference cells, we investigated the ALD-grown TiO_2 underlayer in more detail for the solar cells with P2. The hole blocking or current rectifying properties of the TiO_2 layers have been investigated as a function of their thickness, shown in the next paragraph. Diode structures have been prepared, which consisted only of the FTO electrode, the flat TiO_2 underlayer, the spiro-MeOTAD layer and the gold counter-electrode.

6.3.5 ALD and Characterization of the TiO_2 Layers

With the ALD technique, it is possible to apply very thin, dense, and crystalline TiO_2 layers. It was investigated, which thickness is necessary for proper hole blocking and hence current rectification properties in combination with spiro-MeOTAD. If thinner layers show also good hole blocking properties they might be better suited for solar cells if their oxygen deficiency could be reduced by post-annealing within oxygen. Beside the formation of a Schottky barrier thick TiO_2 layers additionally absorb more of the incident light and might increase the series resistance.

A series of different TiO_2 layer thicknesses has been grown onto FTO and imaged by SEM Fig. 6.10.

An extract of the gas pressure over time plot, recorded during the ALD process is shown in Fig. 6.11.

As can be seen the pressure pulses of the precursor and the water were of a regular height, even after hundreds of cycles. The water pulse was provided in excess to facilitate a complete reaction without a residue of carbon within the TiO_2 layer. The rotary vane pump was continuously evaporating the reaction chamber, decreasing the pressure after each precursor pulse.

Preparation details of the diodes with the composition: FTO/TiO_2/spiro-MeOTAD/gold:

The FTO substrates were pre-treated within a 50-mM aqueous solution of $TiCl_4$ at 70 °C for 1.5 h prior to the TiO_2 layer growth by ALD, as this has been found to improve the statistics and reproducibility of the diodes. After the ALD process, the layers have been annealed within pure O_2 at 500 °C for 2 h. Then a spiro-MeOTAD film has been spin-coated onto each sample at 2000 rpm for 30 s. Afterward, a 50 nm thick gold layer has been sputtered at 0.04 mbar argon through a mask to form eight diodes with 7 mm^2 active area, each. The diodes have been contacted by the golden spring probes of the contacting device. The spiro-MeOTAD solution

Fig. 6.10 SEM micrographs of compact TiO_2 layers grown by ALD. SEM micrographs of an ALD growth series of TiO_2 on FTO substrates. The growth rate was according to the literature [6] 0.03 nm/cycle from the precursor titanium isopropoxide and water at a substrate temperature of 300 °C. Pulse times were 0.6 s, each. The Ti-isopropoxide was evaporated at 80 °C and the water at 47 °C with a 5 s purge by 20 sccm N_2 in between each pulse, resulting in a duration of 11.2 s for a complete cycle

had the composition: 0.15 M spiro-MeOTAD, 0.13 M *tert*-butylpyridine, 9.3 mM TFSI-Li and 0.27 M acetonitrile (for the pre-dissolved TFSI-Li) in chlorobenzene.

The I-V characteristics of the diodes with 60 nm thick TiO_2 layers are shown in Fig. 6.12.

The I-V curves of the diodes on the FTO substrate scattered significantly around a mean value of the series resistance and breakthrough potential (Fig. 6.12a). Upon illumination of the diodes with the lamp, however, the series resistances decreased to less then half of the initial values. Additionally, the opening potentials and

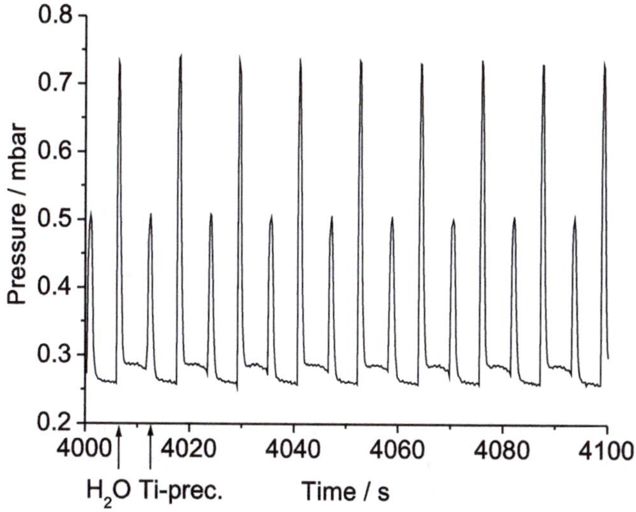

Fig. 6.11 Pressure over time plot recorded during the ALD process. This graph shows the pressure peaks which arose after alternating openings of the titanium precursor (small peaks, Ti-isopropoxide at 80 °C) container valve and the H_2O container valve at 47 °C (large peaks) for 0.6 s each, with 5 s purging time in between each pulse with 20 sccm N_2 permanent flow. The substrate temperature was 300 °C

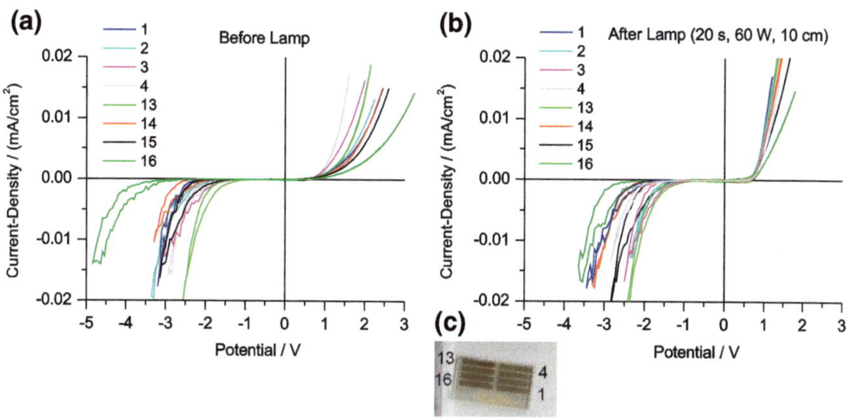

Fig. 6.12 I-V curves of diodes with 60 nm TiO_2 before and after light treatment. The I-V curves of all diodes on a substrate with a 60 nm thick TiO_2 layer and spiro-MeOTAD are shown as prepared (**a**) and after an illumination treatment by the lamp at 10 cm distance for 20 s, which has been repeated up to four times (**b**). The TiO_2 layer was spin-coated by a double layer of spiro-MeOTAD followed by the sputtering of a 51 nm thick gold layer forming eight identical diodes, which were numbered as indicated in (**c**) from 1 to 4 and from 13 to 16. Scan rate: 0.5 V/s. FTO/$TiCl_4$·aq (70 °C)/60 nm ALD-TiO_2/2 h 500 °C in O_2/2 × spiro-MeOTAD (+ TFSI-Li, + tBP)/50 nm Au sputtered/[conditioning in light (lamp)]

series resistances in quadrant I became much more uniform for all diodes on a substrate (b), which is shown in (c). The series resistances of these diodes, which were spin-coated twice by spiro-MeOTAD, decreased from about 60–30 kΩ cm^2. Whereas the breakdown voltages still lay within a broad potential window between approx. −1.5 and −3 V (quadrant III) the forward voltage thresholds (quadrant I) became very narrow distributed around 0.7 V. The major part of the resistance decrease was reached after 20 s of illumination by this warm light. This decrease was irreversible and could be seen also 1 week later, after remounting and re-contacting the diodes at different contact positions. Hence, it was not a contact effect due to a possible softening of the spiro-MeOTAD. When the samples were treated again by the illumination, a further slight decrease of the resistance occurred only temporarily during the illumination period and decayed after about 2 min again to the stable value at the beginning.

The other diodes of the TiO_2 layer thickness series are shown in Fig. 6.13.

These diodes were spin-coated by a single layer of the spiro-MeOTAD matrix and are compared also with a representative of the previously shown diodes with a double layer spiro-MeOTAD (magenta trace). The decrease of the resistance by

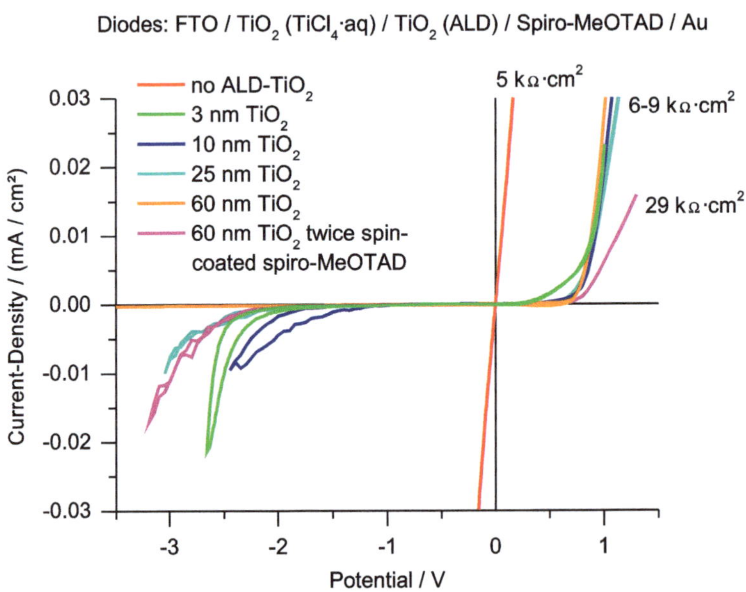

Fig. 6.13 I-V curves of TiO_2/spiro-MeOTAD diodes with different TiO_2 layer thicknesses. The diodes with different TiO_2 layer thicknesses were illuminated by the 60 W lamp at 10-cm distance for a few tens of seconds and measured then in the dark. The FTO layers were pre-treated in a 0.05 M aqueous solution of $TiCl_4$ and then coated with various thicknesses of TiO_2 by ALD from Ti-iProp. They were then spin-coated with one layer of spiro-MeOTAD and sputtered with 50 nm thick gold counter electrodes. The magenta curve, taken from Fig. 6.12, contains a double layer of spiro-MeOTAD. Scan rate: 0.5 V/s. FTO/$TiCl_4$·aq (70 °C)/TiO_2 (ALD)/O_2 (500 °C)/spiro-MeOTAD (+ TFSI-Li, tBP)/50 nm Au sputtered/conditioning in light (lamp)

the illumination treatment reached similar stable values around 8 kΩ cm^2 for all TiO_2 layer thicknesses (Fig. 6.13). The higher series resistance of the diode with the double layer spiro-MeOTAD indicates that this resistance corresponds to the spiro-MeOTAD matrix and is not significantly influenced by the TiO_2 layer thickness. The diode with the 3 nm thick TiO_2 layer grown by ALD (green) has a less defined forward potential and the diode with no TiO_2 layer grown by ALD (red) shows an almost ohmic contact, but with a similar high series resistance. The pre-treatment of all substrates within the aqueous $TiCl_4$ solution is supposed to lead to an initial growth of a 1–2 nm thick TiO_2 layer onto the FTO. However, on the red curve it can be seen, that this pretreatment does not lead to the formation of an effective hole blocking layer.

The breakdown voltage was less affected by the illumination treatment and scattered strongly around an average value also between the diodes of the same substrate. The forward potential instead, became very narrow distributed around 0.7 V for all TiO_2 layer thicknesses from 10 nm onward, after the illumination treatment. The broad distribution of the initial resistance values was reduced to approx. 6–9 kΩ cm^2 upon the irradiation treatment. The mean resistance of the spiro-MeOTAD layer on the FTO without the additional ALD-TiO_2 layer decreased after the illumination treatment only from 5.7 to 5.3 kΩ·cm^2, which corresponds to 8 %. At the 3 nm thick TiO_2 layer the mean decrease was already 72 % and at the TiO_2 layer thicknesses of 10, 25, and 60 nm the mean resistance decrease was about 370, 325, and 50 %, respectively. This shows that the illumination treatment induced the formation of a similar electronic configuration for all TiO_2 layer thicknesses.

The narrowing of the series resistance distribution upon the illumination treatment might point onto a possible reorganization or crystallization of the spiro-MeOTAD matrix. However, it could be seen, that the TiO_2 layer thickness from 10 nm onward had no significant effect on the I-V characteristics. Thinner TiO_2 layers than 10 nm might be amorphous and therefore showed a slightly worse current rectifying characteristic.

The oxygen vacancy doping of the TiO_2 layers could be seen also at their higher absorption within the visible wavelength range. They appeared gray to the naked eye in contrast to the colorless or white appearance of the undoped TiO_2 particles. These layers could not be significantly oxidized upon annealing at 500 °C in pure oxygen for 1 h, for none of the layer thicknesses. The reason for this may be the same as for the observation of Ritala et al. that the crystallinity of ALD-grown layers can not be increased significantly by post-annealing in contrast to CVD–grown TiO_2 films. The authors explained it by a higher density of the ALD-grown films compared to those grown by CVD, what was investigated among other techniques by measuring the etching rate of these layers [6]. An annealing far beyond 500 °C may lead to a coarsening of the crystallites or a phase transformation to rutile from the low temperature modification anatase. The rutile structure instead, is known to perform worse in DSSCs. However, the high oxygen vacancy concentration can also not be avoided during the growth process due to the vacuum, which is known to lead to an oxygen deficiency. Recently, a new ALD system has been developed, which can be operated also under ambient pressure, the so-called spacial ALD. There the gas pulses are not sequentially cycled in time, but the gases are spatially separated and the substrate is oscillating relative to them. Levy et al. demonstrated the growth of Al_2O_3 and ZnO films for thin film transistors employing an open atmospheric pressure system of a spatial ALD [19]. The quality of the films was comparable to those grown by conventional vacuum-based ALD systems. This technique may be suited to deposit high quality low doped TiO_2 layers when pure oxygen would be used as carrier gas. Such layers might perform even better than those prepared by the conventional SPD with oxygen as carrier gas, because they could be prepared more compact and thinner (10 nm), than the SPD–grown once, which need a thickness of about 100 nm to be tight enough for proper current rectification properties.

However, the I-V characteristics of DASCs with P2 under illumination with the lamp seemed to be less affected by the high doping concentration of the TiO_2 layers grown by the vacuum-based ALD system. A kink could also not be observed in conventional DSSCs with the liquid electrolyte and the ruthenium dyes N3 and N719. We therefore assume that the Fermi level of TiO_2 may gets pinned to the redox potential of the I^-/I_3^- couple of the electrolyte. Organic semiconductors are also known to pin the Fermi level of a conductive substrate to the polaronic energy of the organic material [20]. Thus, the DASCs might be less affected by the doping level of our TiO_2 underlayer, under illumination with the lamp, due to a similar Fermi-level pinning to the polaronic level of P2. Snaith et al. showed that a slight kink remains for conventional ss-DSSCs when they are illuminated by monochromatic UV-free light, even when the TiO_2 underlayer has been prepared by the SPD technique under pure oxygen [11]. As P2 aggregate solar cells proved already under the 60 W incandescent lamp that they are less affected by the formation of such a kink in the I-V characteristic under this weak and UV-deficient indoor illumination conditions (Fig. 6.5), they might be better suited for indoor applications than conventional ss-DSSCs.

6.3.6 P2 Aggregate Solar Cells with Treatment in n-heptane

The functionality of the amorphous P2 as sensitizer in the solar cell architecture of ss-DSSCs could be already proved within a mesoporous TiO_2 layer (Fig. 6.5). The following investigations show, if the performance of amorphous P2 could be improved by making use of its biomimetic self-assembly to J-aggregates. The methods how to obtain suitable J-aggregates have been developed in the first part of the thesis, where the best results have been achieved annealing an amorphous P2 layer on flat or within coarse-porous TiO_2 layers in the presence of *n*-heptane. Absorption spectra revealed that this reorganization to J-aggregates was not possible within mesoporous layers (Fig. 47), not even when they were equipped with additional macropores (Fig. 39).

For the following experiments, coarse-porous TiO_2 layers have been applied only onto a part of an TiO_2 coated FTO substrate in order to obtain a partly flat and partly coarse-porous TiO_2 layer on the same substrate, which could be addressed individually by the separated gold electrodes, sputtered on top. This would allow for a better comparison of the performance of P2 aggregates on the flat and within the coarse-porous TiO_2 layer, as the deposition of P2 and its thermal post-treatment in *n*-heptane could be kept identical. Two such substrates have been prepared of which one was treated in *n*-heptane after the deposition of the amorphous P2. Thus, the four solar cell types could be compared: P2 in its amorphous form and as J-aggregates on flat and within coarse-porous TiO_2 layers.

6.3.6.1 Preparation Details

Two FTO substrates were coated with a 25 nm thick TiO_2 layer by ALD, each, with subsequent post-annealing at 500 °C in pure oxygen for 1 h. The coarse-porous TiO_2 (T300) layers were applied by doctor-blading onto one part of each substrate followed by sintering for 1 h at 500 °C in air. Then a treatment within an aqueous solution of $TiCl_4$ (0.05 M) was performed at 80 °C for 3 h. On both substrates 100 μl of a 0.1 mM solution of P2 within THF was sprayed. One of them was annealed within *n*-heptane at 110 °C for 4 h. Then both substrates were first sprayed with the same spiro-MeOTAD solution, which was used for the 3rd N719 cell (0.17 M spiro-MeOTAD, 13 mM TFSI-Li, 0.12 M tBP, 0.42 M acetonitrile) and afterward spin-coated with it at 3000 rpm for 30 s. Finally, 50 nm thick and 7-mm^2 large gold electrodes have been sputtered onto the flat and coarse-porous TiO_2 areas on each substrate (the FTO underneath the top contact area was not etched away).

Fig. 6.14 shows a photo of the *n*-heptane treated substrate and the I-V curves of its highest efficient coarse-porous and flat solar cell under illumination with the lamp (measured 7 months after preparation).

The coarse-porous dye aggregate solar cell reached a ten times higher short circuit current (0.02 mA/cm^2) than the flat one. Its open circuit potential however, was significantly lower and additionally degraded from 0.34 V (not shown) down to 0.28 V after the storage for half a year in the dark (Fig. 6.14a). The V_{oc} of the flat cell did not degrade during this time and remained at 0.63 V, which will be shown later. The reason for the lower V_{oc} of the coarse-porous cell might be a hindered electron transport through the large TiO_2 particles with diameters mainly between 100 and 300 nm, as this size already reached the dimension of a band bending region. In contrast to the N719 reference ss-DSSCs with a mesoporous TiO_2 layer the I-V characteristics of these solar cells did not reveal a significant kink under illumination with the lamp, as the fill-factor was higher than 25 %, what is the threshold for a linear curve progression (triangle) in quadrant IV. The corresponding solar cells are marked by arrows in Fig. 6.14b. It made no difference over which of the two silver paint spots (top and bottom) the FTO electrode was contacted.

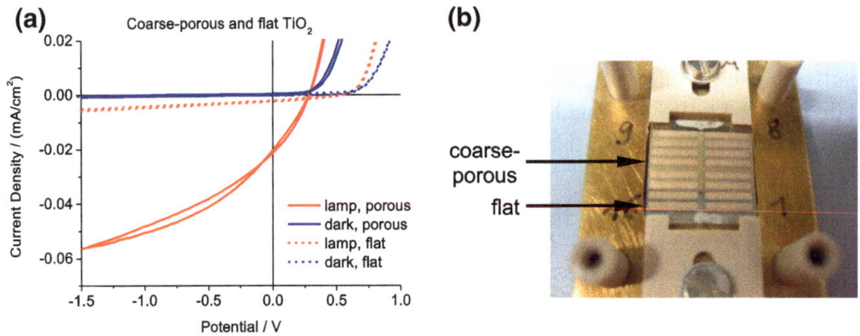

Fig. 6.14 I-V curves of *n*-heptane treated coarse-porous and flat DASC with a photo. **a** Comparison of the I-V curves of the *n*-heptane treated coarse-porous (solid traces) and flat (*dashed traces*) dye aggregate solar cells (DASCs) under illumination with the lamp (60 W, 10 cm distance) 7 months after preparation. Marek PL, Hahn H, Balaban TS (2011) Energy Environ Sci 4:2366-2378 – Reproduced by permission of The Royal Society of Chemistry. **b** Photo of the substrate with arrows indicating the corresponding solar cells. FTO/ALD-TiO_2 (25 nm)/1 h 500 °C in O_2/[coarse-porous TiO_2 (T300)]/1 h 500 °C/$TiCl_4$/1 h 500 °C P2-THF sprayed/*n*-heptane post-treatment/spiro-MeOTAD sprayed/spiro-MeOTAD spin-coated/50 nm Au sputtered/stored 7 months

In Fig. 6.15 the effect of the *n*-heptane treatment is shown in a comparison of the same solar cells with those from the substrate, which has not been annealed within *n*-heptane prior to the spiro-MeOTAD deposition, while the rest of the preparation was kept identical.

In these earlier measurements of the same *n*-heptane treated cells (solid traces) in comparison with the two highest efficient cells on the reference substrate, which has not been annealed in *n*-heptane, (dashed traces) it can be seen, that the *n*-heptane treatment indeed led to a significant increase of the photocurrent. The power output and hence efficiency increased immensely by a factor of 12 on the coarse-porous part (a) and significantly by 30 % aside this structure on the flat TiO_2 layer (b). Despite the lower open circuit potential of the porous cell its efficiency was seven times higher compared to that of the flat cell on the *n*-heptane treated substrate. The possible reasons are discussed at the end of this chapter.

This experiment proved that P2 indeed performs better in form of J-aggregates than in the amorphous phase, especially within the coarse-porous layer.

Fig. 6.16 shows the performance and stability of the *n*-heptane treated coarse-porous DASC under simulated sunlight.

Under simulated sunlight a kink arose also in the I-V characteristic of the 7-month-old coarse-porous DASC (a, green trace). Hence, the overall efficiency was with 0.006 % in the same order of magnitude as that of the N719 reference ss-DSSCs (efficiency of the 3rd N719: 0.008 %). Considering the high photocurrent at –1 V bias it can be estimated that the efficiency could be around two orders of magnitude higher without the kink, similar to the 3rd N719 reference cell. The corresponding I-V curve in the dark is shown as dark green trace. After three days, the same solar cell

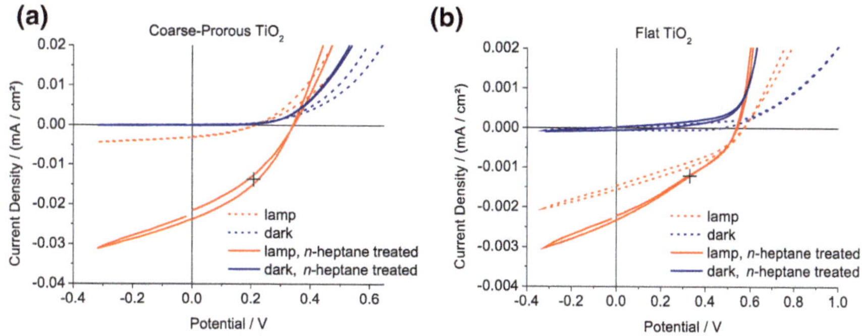

Fig. 6.15 I-V curves of flat and coarse-porous P2-DASCs with and without *n*–heptane post-treatment. I-V curves of the same DASCs as in Fig. 6.14 (solid traces) compared with identically prepared DASCs, but without the *n*-heptane treatment of the amorphous P2 deposits, (*dashed traces*) on the coarse-porous TiO_2 layer (**a**) and on the bare TiO_2 underlayer grown by ALD (**b**). All I-V curves were measured 1 month after solar cell preparation under the lamp (*red* traces) and in the dark (*blue* traces). Note the ten times higher current density scale in **b**. The crosses mark the maximum power points of the *n*-heptane treated solar cells, corresponding to 2.9 $\mu W/cm^2$ and 0.4 $\mu W/cm^2$ in (**a**) and (**b**), respectively. Scan rate: 0.05 V/s. FTO/ALD-TiO_2 (25 nm)/1 h 500 °C in O_2/[coarse-porous TiO_2 (T300)]/1 h 500 °C/$TiCl_4$/1 h 500 °C/P2-THF sprayed/[*n*-heptane post-treatment]/spiro-MeOTAD sprayed/spiro-MeOTAD spin-coated/50 nm Au sputtered/stored 1 month

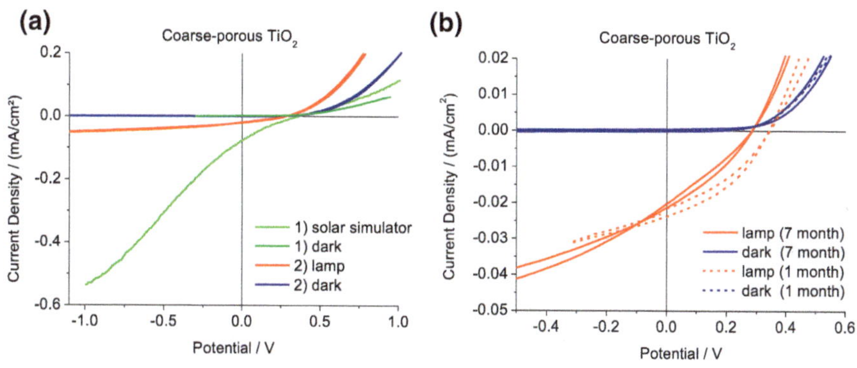

Fig. 6.16 I-V curves of the *n*-heptane treated coarse-porous cell (solar simulator and stability). I-V curves of the same *n*-heptane treated coarse-porous solar cell as shown before in Fig. 6.14 and 82a. **a** Measured 7 months after preparation in the dark (*blue* traces), under illumination with the lamp (*red* trace) and in a solar simulator (green trace, 100 mW/cm^2 AM1.5). Copyright (2011) by The Royal Society of Chemistry. Marek PL, Hahn H, Balaban TS (2011) Energy Environ Sci 4:2366-2378 – Reproduced by permission of The Royal Society of Chemistry. **b** Measured under illumination with the lamp 1 month (dashed *red* trace) and 7 months after preparation and after the measurements in the solar simulator (solid *red* trace) with the corresponding measurements in the dark (*blue* traces)

has been measured again under illumination with the lamp (red trace) with the corresponding dark curve (solid *blue* trace). b) The same I-V curve under illumination with the lamp (solid red trace) is compared with a measurement, which has been recorded 1 month after preparation of this solar cell (dashed red trace) with the corresponding dark measurements (dashed *blue* trace). The open circuit potential and the fill-factor decreased during this half year storage in the dark and slightly also during the solar simulator measurements. The photocurrent, however, did not decrease significantly in contrast to the aged N719 solar cells.

Fig. 6.17 shows a similar comparison for the *n*-heptane treated flat DASC.

The DASC on the flat TiO_2 layer appeared to be less affected by the formation of a kink in the I-V curve, if at all, under simulated sunlight (a, green trace). This may be also related to the generally lower photocurrent. However, its high fill-factor overcompensated the deficit in photocurrent and led to a power output of 9 μW/cm^2 under simulated sunlight, corresponding to 0.009 % power conversion efficiency. Its power output under illumination with the lamp (red trace) was only 0.4 μW/cm^2. However, this flat cell was remarkable stable and showed no degradation after half a year storage in the dark (b, solid red trace) and even performed slightly better than half a year earlier (dashed red trace). As the prolonged storage of the reference cells with the sensitizer N719 always led to an expansion of the kink in the I-V curve and hence decrease of the fill-factor, the flat DASC was inferably not affected by such a kink or the formation of a Schottky barrier. The shunt resistance of this cell decreased significantly under simulated sunlight down to 26 kΩ·cm^2. This indicates high recombination losses, probably at the highly

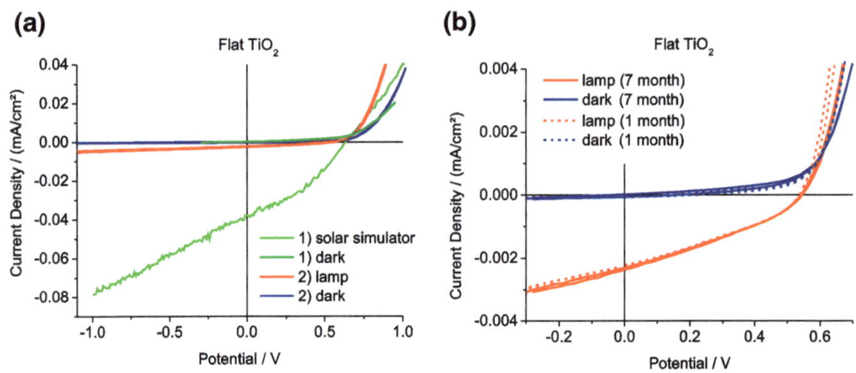

Fig. 6.17 I-V curves of the *n*-heptane treated flat cell (solar simulator and stability). I-V curves of the same *n*-heptane treated flat solar cell as shown before (Fig. 6.14, 6.15b) **a** measured 7 months after preparation in the dark (*blue* traces), under illumination with the lamp (*red* trace) and in a solar simulator (green trace, 100 mW/cm^2 AM1.5). Copyright (2011) by The Royal Society of Chemistry. Marek PL, Hahn H, Balaban TS (2011) Energy Environ Sci 4:2366–2378. **b** Measured under illumination with the lamp 1 month (dashed *red* trace) and 7 months after preparation and after illumination with the solar simulator (solid *red* trace) with the corresponding dark measurements (*blue*)

doped ALD grown TiO_2 layer and the spiro-MeOTAD matrix, which had much direct contact area, as the J-aggregates do not form a closed layer.

The most important electrical parameters of the demonstrated DASCs are compared in Table 3 in the next chapter (Table 7.1).

References

1. Schmidt-Mende L, Grätzel M (2006) Thin Solid Films 500:296–301
2. Snaith HJ, Humphry-Baker R, Chen P, Cesar I, Zakeeruddin SM, Grätzel M (2008) Nanotechnology 19(424003):1–12
3. Fabregat-Santiago F, Bisquert J, Palomares E, Haque SA, Durrant JR (2006) J Appl Phys 100:034510
4. Law M, Greene LE, Radenovic A, Kuykendall T, Liphardt J, Yang P (2006) J Phys Chem B 110:22652–22663
5. Rahtu A, Ritala M (2002) Chem Vap Depos 8:21–28
6. Ritala M, Leskeli M (1993) Chem Mater 5:1174–1181
7. Aarik J, Aidla A, Kiisler A-A, Uustare T, Sammelselg V (1997) Thin Solid Films 305:270–273
8. Burke A, Ito S, Snaith H, Bach U, Kwiatkowski J, Grätzel M (2008) Nano Lett 8:977–981
9. Marek PL, Sieger H, Scherer T, Hahn H, Balaban TS (2009) J Nanosci Nanotechnol 9:3708–3713
10. Hore S, Nitz P, Vetter C, Prahl C, Niggemann M, Kern R (2005) Chem Commun 15:2011–2013
11. Snaith HJ, Grätzel M (2006) Adv Mater 18:1910–1914
12. Kron G, Egerter T, Werner JH, Rau U (2003) J Phys Chem B 107:3556–3564
13. Rühle S, Cahen D (2004) J Phys Chem B 108:17946–17951
14. Dittrich Th, Weidmann J, Koch F, Uhlendorf I, Lauermann I (1999) Appl Phys Lett 75:3980–3982
15. Barnes PRF, Anderson AY, Koops SE, Durrant JR, O'Regan BC (2009) J Phys Chem C 113:1126–1136
16. Ito S, Liska P, Comte P, Charvet R, Péchy P, Bach U, Schmidt-Mende L, Zakeeruddin SM, Kay A, Grätzel M, Nazeeruddin MK (2005) Chem Commun 34:4351–4353
17. Glatthaar M, Riede M, Keegan N, Sylvester-Hvid K, Zimmermann B, Niggemann M, Hinsch A, Gombert A (2007) Sol Energ Mat Sol Cells 91:390–393
18. Wang M, Moon S-J, Xu M, Chittibabu K, Wang P, Cevey-Ha N-L, Humphry-Baker R, Zakeeruddin SM, Grätzel M (2010) Small 6:319–324
19. Levy DH, Freeman D, Nelson SF, Cowdery-Corvan PJ, Irving LM (2008) Appl Phys Lett 92:192101
20. Tengstedt C, Osikowicz W, Salaneck WR, Parker ID, Hsu C-H, Fahlman M (2006) Appl Phys Lett 88(053502):1–3

Chapter 7
Results and Discussion: Application of BLH to Solar Cells

The following table summarizes the electrical parameters of the four dye aggregate solar cells (DASCs): the flat (F) and the coarse-porous (CP) one, with (n-C_7) and without the n-heptane treatment, under illumination with the lamp. The n-heptane treated ones are additionally shown under illumination with simulated sunlight (SS) (Table 7.1).

The high series resistance of the flat solar cell without the n-heptane treatment of 58.3 kΩ cm^2 confirms the assumption that the amorphous P2 might have a significantly higher resistance, compared to the orderly structured P2 J-aggregates. By the treatment with n-heptane the amorphous P2 layer reorganized to J-aggregates, leading to a significant reduction of the series resistance down to 2.2 kΩ cm^2. The series resistance of the coarse-porous solar cell was already low before the n-heptane treatment, probably due to a much thinner amorphous P2 layer on the larger surface area. The treatment of the coarse-porous layer in n-heptane led a reduction of the series resistance from 1.6 to 1.0 kΩ cm^2 under the illumination with the lamp.

The reorganization of the amorphous P2 to J-aggregates in n-heptane led to an impressive increase of I_{SC} by a factor of 7 on the coarse-porous TiO_2, whereas the I_{SC} increase of the flat cell was 50 %. One reason for the smaller effect on the flat TiO_2 surface might be that the aggregates were separated to much from each other and a major fraction of the incident light could pass beside the aggregates without being absorbed. Another explanation might be that the J-aggregates grew larger on the flat surface, exceeding the exciton diffusion length. Another effect, which might explain this observation is the higher light scattering of the coarse-porous layer compared to the flat one. J-aggregates of P2 have additional red-shifted absorption bands, compared to amorphous P2. Whereas this additional gain in harvesting photons may be small on a flat surface, it can be amplified by light scattering, which is much higher in the coarse-porous cell. By the additional absorption bands a certain number of photons more could be harvested for each time the incident light passes the layer. Whereas the light passes a flat layer most probably only once, it may pass it several times within a porous layer, thus multiplying the additional wavelengths gain of J-aggregates. As a relatively low

P. L. Marek, *Biomimetic Dye Aggregate Solar Cells*,
Springer Theses, DOI: 10.1007/978-3-319-00636-9_7,

Table 7.1 Electrical parameters of dye aggregate solar cells (DASC)

DASC	Illumination	P (μW/cm^2)	V_{MPP} (V)	I_{MPP} (μA/cm^2)	FF (%)	V_{oc} (V)	I_{SC} (μA/cm^2)	R_s (+1 V) (kΩ cm^2)	R_{sh} (kΩ cm^2)
CP	Lamp	0.2	0.12	1.9	33	0.22	3	1.6	288
CP, n-C_7	Lamp	1.8	0.16	11	30	0.28	21	1.0	108
CP, n-C_7	SS	6	0.17	33	20	0.36	77	3.5	–
F	Lamp	0.3	0.38	0.8	36	0.57	1.5	58.3	619
F, n-C_7	Lamp	0.4	0.33	1.2	32	0.54	2.3	2.2	467
F, n-C_7	SS	9	0.37	24	37	0.63	38	7.7	26

CP coarse-porous, *F* flat, *SS* solar simulator (100 mW/cm^2 AM 1.5), n-C_7 n-heptane treatment
The values of the n-heptane treated cells correspond to the 7-month-old cells and the others to 1-month-old cells with amorphous P2. The series resistance (R_s) was recorded at +1 V for all cells

concentration of P2 has been used for these cells a high amplification of the extra absorbance of J-aggregates might have occurred.

The reason for the lower open circuit potential of the coarse-porous cell compared to the flat one, may be a band bending within the TiO_2 microparticles, leading to a more rough potential landscape with potential barriers at the particle interfaces. Small nanoparticles instead, are smaller than the size of a depletion layer or band banding zone, thus are almost completely depleted of conduction band electrons and have a flat band potential. Thus, the barrier for the electron transport between such connected particles is supposed to be lower, then that between larger particles [1]. Therefore, a smaller particle size, which would still allow the formation of J-aggregates might lead to a higher open circuit potential.

Conventional porphyrin dyes, which form a monolayer on mesoporous TiO_2 surfaces, are known to have generally a lower open circuit potential than the ruthenium dye N719 [2]. Mozer et al. investigated the electron lifetime and diffusion within such porphyrin stained DSSCs and came to the conclusion, that an increased recombination rate is mainly responsible for the decreased open circuit potential of these porphyrin dye monolayers. The recombination within our dye sensitized solar cells is also relatively high, as the shunt resistances are relatively low. This is probably caused by the exposed TiO_2 surface, because P2 J-aggregates do not form a dense layer. If this recombination path should turn out to be problematic in future, it might be possible to passivate this free TiO_2 surface by suitable co-adsorbents. Also the use of another, less recombination affected hole transport material might be possible in our solar cells, due to the easier infiltrateability of the coarse-porous structure.

For the DASCs under the illumination with the lamp, no significant kink was observed what can be also seen at the fill-factors, which were not lower than 30 % in contrast to the reference solar cells (ss-DSSCs with N719), where the fill-factors were only for the 1st N719 cell above 20 %. The lower photocurrent can not be the only explanation for this, because the comparison of the 1st N719 cell with the first P2 cell on an identically prepared mesoporous TiO_2 substrate (T37) showed that the cell with P2 reached a higher photocurrent and had no kink in the I-V curve under illumination with the lamp (Fig. 7.1), in contrast to the 1st N719 cell (Fig. 7.2). This may indicate that DASCs with P2 are less affected by the formation of a Schottky barrier than ss-DSSCs based on the common ruthenium bipyridine dye N719 under illumination with an incandescent lamp. As such a lamp represents indoor illumination conditions, having also a low UV content, DASCs may have an advantage over conventional ss-DSSCs indoors. As Snaith and Grätzel showed, also those ss-DSSCs with a dense TiO_2 layer, prepared by the spray pyrolysis deposition in pure oxygen, revealed a slight kink, when illuminated under monochromatic light [3]. Thus, even with the best technique to apply the dense TiO_2 layer it seems to be not possible to fully avoid such a kink formation under UV-free illumination conditions. Maybe the P2-based DASCs could perform better under such conditions, which were simulated here with the 60 W incandescent lamp.

These DASCs are a proof of concept and not fully optimized solar cells, thus there is much room for improvements in future. The highest improvement is expected by a change of the application method for the dense TiO_2 hole blocking layer, to one which allows a higher saturation with oxygen. A common deposition technique to achieve this is the spray pyrolysis deposition with pure oxygen as carrier gas for the TiO_2 precursor. The high photocurrent density at negative bias as well as literature values show that an efficiency increase of the reference solar cells based on the N719 dye by two orders of magnitude is possible with such low doped TiO_2 layers. As the performance of the coarse-porous DASCs was also diminished by such a kink under SS, we expect also for this type of solar cells a high efficiency increase. Further improvements are expected by using a larger amount of P2, as these cells absorbed only about 20 % of the incident light. SEM investigations revealed that there is enough space in the pores to significantly increase the concentration of P2. Alternatively, the layer thickness can be reduced to decrease the series resistance and the recombination rate. The usage of smaller particles within the not investigated range between 37 and around 100 nm might also increase the the open circuit potential and hence the efficiency further.

References

1. Hagfeldt A, Grätzel M (1995) Chem Rev 95:49–68
2. Mozer AJ, Wagner P, Officer DL, Wallace GG, Campbell WM, Miyashita M, Sunaharac K, Mori S (2008) Chem Commun 4741–4743
3. Snaith HJ, Grätzel M (2006) Adv Mater 18:1910–1914

Chapter 8
Conclusion: Application of BLH to Solar Cells

The biomimetic self-assemblies or J-aggregates of P2 could be successfully implemented into solar cells. Within conventional DSSCs comprising the liquid iodide-based electrolyte, P2 performed very poor, most probably due to its partial dissolution within such solvent mixtures. The solid-state version of these solar cells, with a mesoporous TiO_2 layer infiltrated by amorphous P2, performed much better. On flat and in coarse-porous TiO_2 layers, which allowed also the reorganization of the amorphous P2 to J-aggregates, it could be shown, that indeed J-aggregates perform significantly better than the amorphous P2. The efficiency of the coarse-porous cells increased by a factor of 12 due to the induced self-assembly to J-aggregates by the thermal post-treatment in *n*-heptane.

The reference solar cells based on the standard ruthenium dye N719 revealed that the TiO_2 hole blocking layer, which was prepared by atomic layer deposition (ALD), was unsuited for these solar cells. Its high oxygen vacancy concentration most probably led to the formation of a Schottky barrier with the FTO, and thus a kink in the I-V characteristic, which significantly reduced the efficiency of these solar cells. Whereas the coarse-porous dye aggregate solar cells (DASCs) with P2 were also affected by the formation of such a kink under illumination with simulated sunlight, this deficit was significantly less pronounced at the weak and UV-deficient illumination in contrast to the reference solar cells. The indoor illumination conditions we simulated by a 60 W incandescent lamp. The efficiencies of the P2-based coarse-porous DASCs and those of the N719-based reference ss-DSSCs at simulated sunlight were comparable, reaching 0.006 and 0.008 %, respectively. The high photocurrents at negative bias of both types of solar cells let expect an efficiency increase about two orders of magnitude when the kink would be eliminated i.g. by using a low doped TiO_2 underlayer. To realize such a TiO_2 layer a new deposition system would be needed to set up in future like the conventional spray pyrolysis deposition or a spatial ALD system, which works also under atmospheric pressure and could be used with pure oxygen, too.

That the DASC with P2 was less affected by the formation of a kink in the I-V characteristic under almost UV-free illumination conditions, compared to the reference ss-DSSCs based on N719, was quite unexpected. This may be an

P. L. Marek, *Biomimetic Dye Aggregate Solar Cells*,
Springer Theses, DOI: 10.1007/978-3-319-00636-9_8,

additional advantage of DASCs for indoor applications and low light intensities. Whereas the N719 reference cells showed a pronounced kink in the I-V characteristics also at very low photocurrents, this kink was not noticeable in identically prepared solar cells with amorphous P2 as sensitizer of the mesoporous TiO_2. This is not only related to a lower photocurrent as the example of the 1st N719 ss-DSSC showed. The reference cell was only weakly stained by N719 due to an intense purging within pure acetonitrile, and thus showed a lower photocurrent than the P2 infiltrated analogue (Figs. 6.5 and 6.6, respectively). Higher photocurrents of the N719 cells could be achieved by purging the cells in an acetonitrile/*t*-butanol mixture instead of pure acetonitrile after staining, treating the TiO_2 layers within aqueous $TiCl_4$ solutions and using smaller TiO_2 particles (13 nm). The strong kink in the I–V curve however, could not be reduced significantly, what was tried with an intensified post-annealing within pure oxygen (Fig. 6.16).

J-aggregates of P2 could be successfully implemented into solar cells using the basic architecture of ss-DSSCs, but in a coarse-porous variant. This resulted in a combination of the features of hybrid solar cells, and ss-DSSCs. The resulting DASCs combine the ability of the dye to transport excitons and holes, like in hybrid solar cells, with the peculiarity of ss-DSSCs to make use of an additional material dedicated to the hole transport only. This concepts offers several advantages compared to conventional ss-DSSCs:

- increased absorbance and sensitization on a smaller TiO_2 surface
  - possibility using thinner layers

    could reduce the series resistance
    higher degree of mechanical flexibility
    savings in material cost and weight

  - possibility of coarsening the porous structure

    reduction of the recombination rate at the TiO_2 surface (due to reduction of the TiO_2 surface)
    facilitates the infiltration of the HTM for a higher pore filling

    larger variety of possible HTM alternatives like those with a higher viscosity

- seemingly better suited for low UV or indoor illumination conditions
- environmentally friendly and not containing rare, expensive or potentially toxic elements.

Possible disadvantages seem to be the lower efficiency up to now, but this is probably related to the early stage of optimization. Beside the doping level of the TiO_2 hole-blocking layer the morphology of the porous layer and that of the dye aggregates has a big influence on the efficiency and may be further optimized, as well as the infiltrated dye amount. A principle disadvantage may be a hindered electron transport through a coarse-porous TiO_2 layer, what might have caused in our case the lower open circuit potential. As nanoparticles are smaller than the depletion layer within the material, their electronic bands are not bended in contrast to microparticles, where the band bending is supposed to form barriers at the TiO_2 particle interfaces. This increases the internal resistance within such coarse-porous solar cells. Thus, a good balance needs to be found between the TiO_2 particle size the formation of voids large enough to allow the self-assembly to J-aggregates.

This work showed that self-assembling dye-aggregates can sensitize semiconductors like TiO_2 for visible light with a similar efficiency as dye monolayers. A proof of principle was given, how self-assembling dye aggregates can be implemented into artificial solar cells. The coarse-porous TiO_2 layer is a suitable template to control the aggregate morphology and size in combination with different solvents or treatments for the infiltration and to induce the self-assembly. The pore sizes should be narrow distributed and on one hand large enough to allow the self-assembly and on the other hand small enough to confine the aggregate size to the exciton diffusion length. The success of the self-assembly to J-aggregates could be monitored by absorption spectroscopy. The distribution of the aggregates' morphologies and sizes could be judged by SEM investigations.

The miniaturized solar cell design improved the statistics and could be used to apply different solar cell structures on a single substrate, like a porous and a flat one. This allowed for a precise comparison of the morphology influence on the performance, as all deposition and post-treatment parameters were identical. With the contacting device, which has been developed to contact precisely the 7 mm^2 small solar cells, it was not necessary to etch away the FTO layer underneath the contact area of the gold counter electrode, because the used spring gold probes did not short circuit the cells. With this switchable contacting device highly reproducible and fast measurements could be realized with more reliable statistics.

Chapter 9
General Conclusion

A biomimetic self-assembling dye, which forms aggregates that mimic the natural light-harvesting system of special photosynthetic active bacteria, has been investigated toward its applicability to solar cells. This fully synthetic dye, self-assembles to orderly structured nano- to micrometer sized rod-shaped aggregates, which might improve solar cells based on conventional organic dyes. In order to use the full potential of the dye aggregates, the self-assembly needed to be controlled and a suitable solar cell concept for their implementation developed. In the first part of this thesis, it has been investigated how the self-assembly can be controlled in order to achieve dye aggregates with a high internal order and size-confinement at least in one dimension. This dimension should not exceed the exciton diffusion length, what is the maximum path length, an excited state, which originates by the absorption of light, can travel within an aggregate. In the second part of this thesis, the findings from the first part have been applied to construct and test first solar cell prototypes with such controlled dye aggregates.

In contrast to the chlorophylls of plants, the self-assembling bacteriochlorophylls of green sulfur bacteria, which served as model for our dye, do not require proteins for the assembly of functional light-harvesting systems. This allows to mimic the light-harvesting system fully synthetically in large scale in order to realize low-cost, lightweight, and environmentally friendly solar cells. As nature gives the proof of principle in practice for the functionality of such dye aggregates, this work has been carried out in order to investigate the transferability of this concept to technology.

Various methods have been investigated in order to decrease the size of the aggregates while maintaining as far as possible their highly ordered internal structure. These included the steric hindering of the self-assembly within voids of different porous TiO_2 layers and on zinc oxide nanorod substrates, the kinetic growth-hindering at low temperatures and a size-selective deposition by dielectrophoretic forces. Other investigated techniques to deposit the aggregates were spin-coating, spraying, precipitation and a two-step process, where at first an amorphous layer was deposited suppressing the self-assembly in order to initiate afterward a reorganization to the desired aggregates. The degree of success was

P. L. Marek, *Biomimetic Dye Aggregate Solar Cells*,
Springer Theses, DOI: 10.1007/978-3-319-00636-9_9,

monitored mainly by absorption spectroscopy in combination with scanning electron microscopy (SEM). Whereas the absorption spectroscopy gave a value of the average degree of the aggregates' internal order, the SEM allowed an assessment of the aggregates' dispersity, that is, the fraction and distribution of different morphologies, sizes, and their alignment on the substrate.

Finally, these aggregates have been implemented into solar cells, designed to combine the advantages of hybrid solar cells and solid-state dye-sensitized solar cells (ss-DSSCs). We call them: dye aggregate solar cells (DASCs). They were constructed on flat and different porous TiO_2 structures, which were sensitized by the dye aggregates for visible light with supplement for the hole transport by the transparent organic hole transport material spiro-MeOTAD. The performance of the DASCs was compared with that of ss-DSSCs based on the common ruthenium dye N719, which is often used as reference. Additional investigations have been done using the atomic layer deposition (ALD) technique to apply dense hole-blocking TiO_2 layers onto the transparent conductive substrates with a fluorine doped tin oxide (FTO) layer. As application method for the gold counter electrode the sputtering technique was used for these solar cells as alternative to the conventional evaporation technique. The development of a special contacting device allowed miniaturized solar cells to be contacted without the need for pre-structuring the commercial FTO-layers.

The biomimetic dye aggregates could be successfully controlled in their size and dispersity and they could be successfully implemented into functional solar cells, giving a proof of principle that the transfer of this natural light-harvesting concept to technology is possible.

The precondition for realizing efficient solar cells based on dye aggregates was to find methods to reduce the aggregate size close to the exciton diffusion length while maintaining the highly ordered J-aggregate structure. This could be achieved, and additionally a homogeneous deposition with a narrow distribution of the different aggregate sizes and morphologies, by a two step process on suitable substrates. Therefor, P2 was first applied from the solvent THF, which hinders the self-assembly, resulting in the deposition of amorphous, but very homogeneous films. A reorganization of these films to J-aggregates could be obtained by a thermal post-treatment within *n*-heptane, a solvent which facilitates the self-assembly without dissolving the dye. A further size-reduction was achieved by applying this method to a coarse-porous TiO_2 structure, where the aggregates grew size-confined within the pores. This reorganization was not possible within mesoporous TiO_2 layers with particle- and hence pore-sizes up to approx. 37 nm. Even, when additional macropores were implemented into such layers, as could be measured by absorption spectroscopy which revealed that a major fraction of the dye remained amorphous within the mesopores and only a small fraction grew to J-aggregates within the macropores. A complete reorganization of the amorphous dye deposits could be achieved within coarse-porous TiO_2 layers, sintered from particles of around 200 nm size. Due to the relatively high aspect ratio of the aggregates the diameter of even 200 nm long aggregates was still within the order of the expected exciton diffusion length. The superiority of self-assembled

J–aggregates over the amorphous dye could be proved by a rise in efficiency of over one order of magnitude upon treatment of the dye-infiltrated substrates in *n*-heptane before finishing them to solar cells. These DASCs had the following structure:

FTO/dense TiO_2/coarse-porous TiO_2/P2 J-aggregates/spiro-MeOTAD/gold

The main factor, which limited the efficiency of the DASCs and that of the reference solar cells (ss-DSSCs), which were sensitized by a monolayer of the highest efficient ruthenium-dye, was a highly doped and hence unsuited TiO_2 hole blocking layer. It most probably led to a Schottky barrier formation and hence kink in the I-V characteristic, which reduced the efficiency by estimated two orders of magnitude. However, the efficiency of the new developed coarse-porous DASC reached with 0.006 % already ¾ of the efficiency of the reference ss-DSSC under simulated sunlight. This shows the high potential of the concept using biomimetic dye aggregates for solar cells as the relatively low optimized DASC prototypes were already compatible with ss-DSSCs comprising conventional dye monolayers.

The following suggestions for optimizations, summarized in the outlook, are expected to lead to further improvements of the DASC performance.

Chapter 10
Outlook

Further optimizations of P2 aggregate solar cells might be possible by:

- using lower doped TiO_2 hole blocking layers (prepared, e.g., by SPD or spatial ALD in O_2 at atmospheric pressure)
- using higher amounts of the dye P2 (approx. up to 1×10^{-7} mol/cm^2)
- using thinner coarse-porous TiO_2 layers (<1 μm)
- adjusting of the porosity, e.g., with TiO_2 particles between 37 and 100 nm in size
- infiltration of other HTMs into the coarse-porous TiO_2 (e.g. PEDOT:PSS).

Proposed further investigations:

- internal and external quantum efficiency measurements (IQE, EQE)
- efficiency measurements at different illumination intensities
- stability tests (accelerated aging under illumination at elevated temperatures)
- upscaling of the cells and combination to modules.

P. L. Marek, *Biomimetic Dye Aggregate Solar Cells*,
Springer Theses, DOI: 10.1007/978-3-319-00636-9_10,

Appendix
Literature Survey of Recipes for DSSCs and ss-DSSCs

P. L. Marek, *Biomimetic Dye Aggregate Solar Cells*,
Springer Theses, DOI: 10.1007/978-3-319-00636-9,

DSSCs with liquid electrolytes			Basic Dye Properties					Solar Cell Characterization												Electron transport														
																				anode									sintering		$TiCl_4$-treatment FTO (TiO_2)			
dye		reference	ε / $M^{-1}cm^{-1}$	at λ / nm	solvent	E_{HOMO} / eV	E_{LUMO} / eV	area / cm^2	Intensity	I_{SC} / (mA/cm^2)	U_{OC} / mV	FF	η / %	IPCE / %	at λ / nm	remaind η / %	at °C	after h	comment	TCO	Ω / sq	Cleaning procedure	e^- cond.	layer / µm	dense/nm	applicat.	particle size/nm	appl.	temp./°C	time/h	conc./mM	temp./°C	time/h	comment
Z907+HDMA	31	Wang2003							100 mW/cm²	15.2	764	0.676	7.8			>90	55	1000		FTO			TiO_2	10 (4)	-		20 (400)	SP	500	0.25	50 (40)	70	0.5	
Z907	31	Wang2003							100 mW/cm²	14.2	713	0.708	7.2							FTO			TiO_2	10 (4)	-		20 (400)	SP	500	0.25	50 (40)	70	0.5	
Z907	19	Kuang2008								13.9	731	0.69	7.0																					
K51	19	Kuang2008								14.8	730	0.715	7.7																					
K68	19	Kuang2008								14.4	762	0.69	7.6																					
K19	22	Nazeeruddin2007							100 mW/cm²	15.1	747	0.699	8.0							FTO		det.(15),H_2O,Et,UV-O_3(18	TiO_2	maybe 12(3)			20(400)		500	0.5	40 (40)	70	0.5	
N719	11	Ito2005	1.74E+04	535	Et.					16.6	778	0.731	9.4							FTO	10		TiO_2	14(4)	-		20(400)	SP			-			
N719	11	Ito2005	1.36E+04	535						17.6	805	0.738	10.5							FTO	10		TiO_2	14(4)	yes		20(400)	SP			-			
N719	11	Ito2005								18.7	798	0.713	10.6							FTO	10		TiO_2	14(4)	yes		20(400)	SP			40	70	0.5	
N719	22	Nazeeruddin2007							100 mW/cm²	15.82	785	0.75	9.31							FTO			TiO_2						500	0.5	40 (40)	70	0.5	
N719	2	Boschloo2006						0.785												FTO	8		TiO_2	13				CP	450	0.5				
N719	2	Boschloo2006						0.785			x + 260									FTO	8		TiO_2	13				CP	450	0.5				
N719	24	Saito2008							(100 mW/cm²)	18.11	621	0.585	6.58										ZnO	16.1 (6.7)			(20-30)							
N719	15	Karthikeyan2007				5.9	-3.4																											
N719	20	Martinson2007						0.28	906 W/m²	3.3	739	0.64	1.6							AZO			ZnO	60			Al_2O_3 template	ALD	400	0.5	-			ALD-ZnO on Al_2O_3
N719	14	Kakiuchi2006							100 mW/cm²	12.6	668	0.481	4.1	60	460-550					ITO	4		ZnO	20			ca. 12		450	0.6				
N719	21	Mozer2008								11.68	775	0.72	6.5							FTO	10		TiO2	4.5			13		550	0.5	-			
N719	13	Jiang2006						0.14	100 mW/cm²	19	0.66	0.70	8.9	max. 80	540					FTO	10		TiO2	12 (4)			20 (400)		430 / 520	0.5	40			
N719	16	Koops2009							1 sun	9.55	670		3.82	APCE 78	467					FTO	15	Ac, Me, Hel.	TiO2	4	-			DB	450		-			
N719	23	O'Regan07						5x0.5cm												FTO	8		TiO_2	3			20	SP	570	10min	50	70	0.5	silver stripes
N3	9	Horiuchi2003	1.39E+04	541	AN:t-butar	-5.58		0.25	100 mW/cm²	16.77	698	0.54	6.3							FTO			TiO_2						550	2				
D102	9	Horiuchi2003	5.58E+04	491	AN:t-butar	-5.56		0.25	100 mW/cm²	17.76	604	0.57	6.1							FTO			TiO_2						550	2				
NKX-2311	9	Horiuchi2003				-5.54			100 mW/cm²	16.1	491	0.41	3.3							FTO			TiO_2											
D21L6	112	Yum2009	3.70E+04	458	Et.	0.98V vs.NHE	-1.35V vs.NHE		100.19 % sun	14.1	728	0.71	7.25			90	60	1000	stab. test in IL				TiO_2											
D5	8	Hagberg2008						0.2	100 mW/cm²	12	688	0.72	5.94	~ 85	~ 450-560					FTO			TiO_2	7 (5)	-		20 (400)	SP	500	0.5	yes (yes)			
D7	8	Hagberg2008						0.2	100 mW/cm²	11	695	0.71	5.43	~ 85	~ 450-550					FTO			TiO_2	7 (5)	-		20 (400)	SP	500	0.5	yes (yes)			
D9	8	Hagberg2008						0.2	100 mW/cm²	14	694	0.71	6.9	~ 85	~ 450-590					FTO			TiO_2	7 (5)	-		20 (400)	SP	500	0.5	yes (yes)			
D11	8	Hagberg2008						0.2	100 mW/cm²	13.5	744	0.7	7.03	~ 85	~ 450-590	90	60	1000		FTO			TiO_2	7 (5)	-		20 (400)	SP	500	0.5	yes (yes)			
D11	8	Hagberg2008						0.2	100 mW/cm²	12.3	765	0.7	6.59	79	~ 450-590					FTO			TiO_2	2.5 (5)	-		20 (400)	SP	500	0.5	yes (yes)			
D149	12	Ito2006	6.87E+04	526						19.96	653	0.694	9.03							FTO			TiO_2	12.6(4-5)	-		(400)	SP	500	0.5	40 (40)	70	0.5	
C201	32	Wang2008							100 mW/cm²	13.35	777	0.749	7.8	>80	~ 450 - 580					FTO			TiO_2	7(4)	-		20 (400)	SP						
(ZnPor)	4	Campbell2007																		FTO	15		TiO_2	10 (4)			20 (400)	SP	500	0.5				
DSSCs with ionic liquids:																																		
D21L6	33	Yum2009							1 sun	13.44	686	0.65	5.98										TiO_2											
C201	32	Wang2008							100 mW/cm²	12.4	723	0.779	7			6.2	60	1000	stability test w/	FTO			TiO_2	7(4)	-		20 (400)	SP						
ss-DSSCs with the solid hole transport material spiro-MeOTAD:																																		
N3	1	Bach1998			Et.				9.4 mW/cm²	0.32	342	0.62	0.74	33		80-120 %		80 h		FTO	10		TiO_2	4.2	yes	SPD		SP						
N719	17	Krüger2001						1.07	100 mW/cm²	5	910		2.56	~37	520	95%	RT	3 month		FTO	10		TiO_2	2.5	yes	SPD								
N719	18	Krüger2002						0.16		3.5	821	0.69	2.1																					
N719	18	Krüger2002						0.16		4.8	931	0.71	3.2																					
N719	6	Fabregat-Santiago2006						0.2					1.0										TiO_2	6	sev. tens		12		450	0.5				
Zn-1	25	Schmidt-Mende2005				-5.62	-3.46		100 mW/cm²	5.0	731	0.663	2.44	20 / 60	560-630/440					FTO	10		TiO_2	~ 2	<100	SPD	18	DB	450					
Zn-2	25	Schmidt-Mende2005				-5.52	-3.56		100 mW/cm²	5.9	790	0.651	3.0	25 / 65	570-650/440					FTO	10		TiO_2	~ 2	<100	SPD	18	DB	450					
D102	27	Schmidt-Mende2005b	5.58E+04					0.152	100 mW/cm²	7.7	866	0.612	4.1	~60	440-550				cheap dye	FTO			TiO_2	1.6	100	SPD	18		450		20			
Z907	26	Schmidt-Mende2005a							100 mW/cm²	8.3	752	0.64	4.0										TiO_2	2	yes	SPD		DB	450		-			
Z907	26	Schmidt-Mende2005a							100 mW/cm²	5.5	858	0.67	3.1										TiO_2	2	yes	SPD		DB	450		-			
Z907	**28**	**Schmidt-Mende2006**							**100 mW/cm²**	**8.32**	**752**	**0.643**	**4.03**							**FTO**		**Ac, Me, Hel.**	**TiO_2**	**2**	**100**	**SPD**	**19**	**DB**	**450**	**0.25**	**20**	**RT**	**o.n.**	
Z907	**7**	**Fabregat-Santiago2009**						**0.128**		**9.1**	**860**	**0.51**	**4**							**FTO**	**10**		**TiO_2**	**1.8**	**100**		**20**							
Z907	5	Ding2009																		FTO	15				100	SPD-O_2		SP	500 (450)	0.5	20 (20)	60	6	
K51	29	Snaith2006a						0.16												FTO	15	O_2-pl.(20)			100	SPD-O_2		DB	500	0.5	20	RT	6	
K68	30	Snaith2008																		FTO	15		TiO_2		100	SPD								
K68	3	Chen2009b								5.7	896.7	0.761	3.88							FTO		Ac, Hel., Et, UV-O_3	TiO_2		100	SPD	30	DB	500 (450)	0.5	20	RT	12	TiO_2 from Showa
N719	10	Howie2007						0.25												FTO	15		TiO_2	2-3	yes	SPD	9		550					
N719	10	Howie2007																																
D21L6	33	Yum2009	3.70E+04	458	Et.	0.98V vs.NHE	-1.35V vs.NHE		99.52 % sun	9.64	798	0.57	4.44	54	460								TiO_2	1.7			20							
N719		This work						0.07	100 mW/cm²	0.13 (1.43 at -1 V)	510	0.12	0.006			3rd N719 cell, 7 months old, kink				FTO	10	Deconex, Et, 1h 500 °C	TiO_2	ca. 3	25	ALD	13	DB	500	2	-	-	-	+ 1 h at 500 °C in O2
P2		This work	4E+05	432	THF	-6.2	-3.2	0.07	100 mW/cm²	0.08 (0.54 at -1 V)	370	0.20	0.006			coarse-porous cell, 7 months old, kink				FTO	10	Deconex, Et, 1h 500 °C	TiO_2	ca. 3	25	ALD	ca. 200	DB	500	1	50	80	3	+ 1 h at 500 °C in O2
D5	8	Hagberg2008							99 % sun	6.31	865	0.57	3.11			90	60	1000		FTO	15	Hel, Ac, Et	TiO_2	1.7	100	SPD	20	DB	500	0.5	20		o.n.	annealed in O_2
D7	8	Hagberg2008							99 % sun	5.02	785	0.71	2.79			90	60	1000		FTO	15	Hel, Ac, Et	TiO_2	1.7	100	SPD	20	DB	500	0.5	20		o.n.	annealed in O_2
D9	8	Hagberg2008							99 % sun	7.72	756	0.56	3.25			90	60	1000		FTO	15	Hel, Ac, Et	TiO_2	1.7	100	SPD	20	DB	500	0.5	20		o.n.	annealed in O_2
D11	8	Hagberg2008							99 % sun	5.85	811	0.63	3.01			90	60	1000		FTO	15	Hel, Ac, Et	TiO_2	1.7	100	SPD	20	DB	500	0.5	20		o.n.	annealed in O_2
D9	8	Hagberg2008							9.2 % sun				3.5	40-50	420-530					FTO	15	Hel, Ac, Et	TiO_2	1.7	100	SPD	20	DB	500	0.5	20		o.n.	annealed in O_2
C201	32	Wang2008	4.12E+04	514	chloroform	1.0 vs. NHE	(-0.78) --> -0.87 vs. NHE		100 mW/cm²	9.08	860	0.61	4.8							FTO			TiO_2	1.7	yes	SPD	20	DB	500		20		o.n.	after TiCl4: 450 °C
Z907	7	Fabregat-Santiago2009																																

Abbreviations of methods:

AA	acetic acid	RT	room temperature
appl.	application	SC	spin coated
CP	compressed	SP	screen printing
DB	doctor bladed	SPD	spray pyrolysis deposition
det.	detergent	SPD-O_2	SPD with O_2 as carrier gas
disp.	dispergent	sput.	sputtered
Et.	ethanol	TCO	transparent conductive oxide
EC	ethyl cellulose	th.	thickness
evap.	evaporated	terp.	terpineol
Hel.	Helmanex detergent	UB	ultrasonic bath
IL	ionic liquid	UV-O_3	UV-ozone
Me.	methanol		
o. n.	over night		
O_2	in O_2 atmosphere		
O_2-pl.	O_2-plasma		

Abbreviations of compounds:

AN	acetonitrile	H2PtCl6	hexachloroplatinic acid	PEG	Polyethylene glycol
AZO	Al:ZnO	HDMA	hexadecylmalonic acid	pmii	1-propyl-3methylimidazolium iodide
bmii	butylmethyl imidazolium iodide	ITO	Sn:In_2O_3	Sb dopant	(tris (4-bromophenyl)ammoniumyl hexachloroantimonate
BN	Benzonitrile	K68	Ru-thiocyanato-bipyridyl complex	Spiro-MeOTAD	2,2',7,7'-tetrakis-(*N*,*N*-di-*p*-methoxyphenyl-amine)9,9'-spirobifluorene
Bu_4N	tetrabutyl-ammonium	Li triflate	Lithium trifluoromethanesulfonate	*t*-butanol	*tert*-Butanol
CB	Chlorobenzene	mbii	*N*-methyl-*N*-butyl imidazolium iodide	TBAI	Tetrabutylammonium iodide
D102	Indoline dye	MP	1-methoxy-2-propanol	tBP	4TBP, 4-*tert*-Butylpyridine
D149	Indoline dye	MPN	3MPN, 3-Methoxypropionitrile	TFSI-Li	Bis(trifluoromethane)sulfonimide lithium salt
dmii	1,3-dimethyl-3-imidazolium iodide	N3	Ru-thiocyanato-bipyridyl complex	THF	tetrahydrofuran
dmpii	1,2-dimethyl-3-propylimidazolium iodide	N621	Ru-thiocyanato-bipyridyl complex	VN	valeronitril
$EMIB(CN)_4$	1-ethyl-3methyl-imidazolium tetracyanoborate	N719	Ru-TBA-thiocyanato-bipyridyl complex	Z-907	Ru-thiocyanato-bipyridyl complex
emii	1-ethyl-3-methyl-imidazolium iodide	N749	Ru-thiocyanato-terpyridyl complex	ZnPor	Zinc porphyrin
FTO	F:SnO_2	NBB	*N*-butyl benzimidazole		
GBA	4-guanidinobutyric acid from Fluka	NKX-2311	Coumarine dye		
GuNCS	guanidinium thiocyanate (gtc)	NMBI	*N*-Methylbenzimidazole		

Dye: Staining dye-conc. / mM	solvent	time/h	temp./°C	comment	Hole transport: solvent 1	Vol.-%	solvent 2	Vol.-%	Iodine (I_2) / M	Spiro-MeOTAD / M	LiI / M	TFSI-Li / mM	TFSI-Li / mol-% spiro.	tBP / M	tBP / mol-% spiro.	Add. 1	conc. / M	Add. 2	conc. / M	sealant	cathode	appl.	th. / nm	comment
0.3 + 0.075	AN:t-butanol 1:1				MPN	100			0.1							**pmii**	0.6	NMBI	0.45	Bynel hot melt ring, 35 µm thick				heating posttreatment at 55°C for 4 h, electrolyte filled in with vacuum, hole drilled by sand blasting
0.3	AN:t-butanol 1:1	12	RT		MPN	100			0.1							**pmii**	0.6	NMBI	0.45	Bynel hot melt ring, 35 µm thick				heating posttreatment at 55°C for 4 h, electrolyte filled in with vacuum, hole drilled by sand blasting
																				35 µm thick Bynel rin	Pt-FTO	0.005 M hexachloroplatinic acid in isopropanol at 400 C for 15 min		
																				35 µm thick Bynel rin	Pt-FTO	0.005 M hexachloroplatinic acid in isopropanol at 400 C for 15 min		
																				35 µm thick Bynel rin	Pt-FTO	0.005 M hexachloroplatinic acid in isopropanol at 400 C for 15 min		
0.3	AN:t-butanol 1:1	20-24	RT		MPN	100			0.1							dmpii	0.8	NMBI	0.5	25 µm Surlyn 1702	Pt-FTO	H2PtCl6 in ethanol (2		hole drilled by sand blaster, then cleaned with water in USB, aceton and 0.1 M HCl-aq, after Pt application: 400°C 15 min
					AN	85	VN	15	0.03					0.5		mbii	0.6	GuNCS	0.1	Surlyn 1702	Pt-FTO	H2PtCl6 SC		
					AN	85	VN	15	0.03					0.5		mbii	0.6	GuNCS	0.1	Surlyn 1702	Pt-FTO			
0.5	AN:t-butanol 1:1	24			AN	85	VN	15	0.03					0.5		mbii	0.6	GuNCS	0.1	Surlyn 1702	Pt-FTO			
0.3	AN:t-butanol 1:1	20-24	RT		MPN	100			0.1							dmpii	0.8	NMBI	0.5	25 µm Surlyn 1702	Pt-FTO	H2PtCl6 in ethanol (2		hole drilled by sand blaster, then cleaned with water in USB, aceton and 0.1 M HCl-aq, after Pt application: 400°C 15 min
0.5	ethanol	12	RT		MPN	100			0.05		0.7			0.055										
0.5	ethanol	12	RT		MPN	100			0.05		0.7			0.5										
0.3	ethanol	1.5	60		MPN	100			0.05		0.1			1		dmpii	0.6							
0.5	ethanol	0.5		desorption in 10 mM KOH	MPN	100			0.05		0.5			0.5						Surlyn	Pt-FTO			
0.3	ethanol	3	80	1h excess-dye desorption																				
0.3	AN:t-butanol 1:1	18			AN	100			0.05		0.1			0.5		dmpii	0.6							
0.3	AN:t-butanol 1:1	20	RT		AN	50	VN	50	0.05		0.1			0.5		pmii	0.6				Pt-FTO	sputtered		
0.3	AN:t-butanol 1:1	o. n.		rinsed with AN	**MPN**	**100**			**0.1**		**0.1**			**0.5**		**TBAI**	**0.6**			Surlyn 1472	Pt-FTO			
					AN	100			0.05		(0.1)			0.5		pmii	0.6	(GuNCS)	(0.1)					Examination the effect of $TiCl_4$ treatment
0.5	AN:t-butanol 1:1	17			MPN	100			0.05		0.1			0.6		dmii	0.5				Pt-Ti			
0.5	AN:t-butanol 1:1	17			MPN	100			0.05		0.1					dmii	0.5				Pt-Ti			
					MPN	100			0.05		0.1					dmii	0.5				Pt-Ti			
0.3	ethanol	15	RT		AN	85	VN	15	0.04		0.025			0.28		mbii	0.6	GuNCS	0.05					
0.3	ethanol+10 mM Cl	4	RT		AN	85	VN	15	0.04		0.025			0.28		mbii	0.6	GuNCS	0.05	Surlyn 1702, 25µm	Pt-FTO			Sealant Bynel used to close the hole, antireflection and UV-cut off-film was attached: ARKTOP from Asahi Glass
0.3	ethanol+10 mM Cl	4	RT		AN	85	VN	15	0.04		0.025			0.28		mbii	0.6	GuNCS	0.05	Surlyn 1702, 25µm	Pt-FTO			
0.3	ethanol+10 mM Cl	4	RT		AN	85	VN	15	0.04		0.025			0.28		mbii	0.6	GuNCS	0.05	Surlyn 1702, 25µm	Pt-FTO			
0.3	ethanol+10 mM Cl	4	RT		AN	85	VN	15	0.04		0.025			0.28		mbii	0.6	GuNCS	0.05	Surlyn 1702, 25µm	Pt-FTO			
0.3	ethanol+10 mM Cl	4	RT		AN	85	VN	15	0.04		0.025			0.28		mbii	0.6	GuNCS	0.05	Surlyn 1702, 25µm	Pt-FTO			
0.5	AN:t-butanol 1:1	4	RT		AN	85	VN	15	0.05		0.1			0.05		bmii	0.6			Bynel, 35 µm	Pt-FTO			
0.3	sat. acid in CB	5	RT		AN	85	VN	15	0.03		0.05			0.5		dmii	1			25 µm Surlyn	Pt-FTO			hole drilling by an ultrafine sandblaster, vacuum backfilling, Bynel for hole sealing
0.2	ethanol	2																			Pt-FTO			0.05 M hexachloroplatinic acid
					pmii	65	EMIB(CN)$_4$	35	0.2							NBB	0.5	GuNCS	0.1					
0.3	sat. acid in CB	5	RT						1.67%							mixture of: dmii, emii, EMIB(CN)$_4$, NBB, GuNCS				25 µm Surlyn	Pt-FTO			hole drilling by an ultrafine sandblaster, vacuum backfilling, Bynel for hole sealing
	AN				CB	95	AN	5		0.17		50	29.4			Sb dopant	3.30E-04	AN	0.96		Au	evap.	semitransparent	
0.5	AN:t-butanol 1:1	8	RT		CB	100				0.17		0.26	0.2	0.18	105.88	Sb dopant	7.60E-04				Au			silver stripes to enhance current collection, sealed with glas and Surlyn and Torr Seal
0.5	AN:t-butanol 1:1			Bu_4N, dcbpyH_2	CB	100				0.1		27	27.0	0.11	110.00	Sb dopant	2.00E-04	VN	0.11		Au			
0.5	AN:t-butanol 1:1			silver nitrate (50 mol-% of c	CB	100				0.1		27	27.0	0.11	110.00	Sb dopant	2.00E-04	VN	0.11		Au			
	ethanol	24		afterw. rinsed w. ethanol	CB	100				0.17		15	8.8	0.11	64.71	-					Au			
0.2	THF	o. n.			CB	100				0.17	Li triflate:	13	7.6	0.13	76.47	Sb dopant	3.00E-04				Au		30	Li triflate inst. Li salt
0.2	THF	o. n.			CB	100				0.17	Li triflate:	13	7.6	0.13	76.47	Sb dopant	3.00E-04				Au		30	Li triflate inst. Li salt
0.5	AN:t-butanol 1:1	1			CB	100						yes		yes		Sb dopant	yes				Au	evap.	30	1 min penetration of Spiro-solution, dried for 30 min in Ar-flow and evacuated at $\sim 5\times10^{-6}$ mbar
0.3	AN:t-butanol 1:1									yes		yes		yes		Sb dopant	yes				Au	evap.	30	
0.3	AN:t-butanol 1:1			silver nitrate						yes		yes		yes		Sb dopant					Au	evap.	30	
0.3	**AN:t-butanol 1:1**		**RT**		**CB**	**100**				**0.17**		**13**	**7.6**	**0.13**	**76.47**	**Sb dopant**	**3.00E-04**				**Au**	**evap.**	**30**	
					CB	**100**				**0.17**		**13**	**7.6**	**0.11**	**64.71**						**Au**			
6	GBA : MP 1:1	10 min	RT	rinsed with AN	CB	100	(AN)	(1.5)				8.7		0.047							Au	evap.	100	
0.3	AN:t-butanol 1:1				CB	100	(AN)	(2.6)		0.15		21.9	14.6	0.12	80.00	Sb dopant	8.60E-03	AN	0.50		Au	evap.	30	evap. at 10^{-6} mbar
0.5	AN:t-butanol 1:1				CB	100	(AN)	(1.5)		0.15		8.7	5.8	0.047	31.33			AN	0.29		Au		30	
					CB	100				0.17		19.5	11.5	0.12	70.59						Au	evap.	30	
1	AN:t-butanol 1:1	12	RT		CB	100				**0.18**		0.21	0.1	0.11	61.11	Sb dopant	2.60E-04				Au			
0.3	ethanol	15	RT		CB	100				yes		yes									Au		50	
0.5	AN:t-butanol 1:1	2h	60	rinsed with AN:t-butanol	CB	100	(AN)	(2.2)		0.17		13	7.6	0.12	80.00	-	-	AN	0.42		Au	sputtered	50	7 months old cell with kink in I-V curve
0.1 (100 µl)	THF	sprayed	RT / (110)	n-heptane post-treatment	CB	100	(AN)	(2.2)		0.17		13	7.6	0.12	80.00	-	-	AN	0.42		Au	sputtered	50	7 months old coarse-porous dye aggregate solar cell (DASC) with kink in I-V curve, n-heptane treatment: 4 h at 110°C
		6		rinsed with AN	CB	100						yes		yes							Au		50	Spiro spincoating like Snaith&Grätzel06_2
		6		rinsed with AN	CB	100						yes		yes							Au		50	Spiro spincoating like Snaith&Grätzel06_2
		6		rinsed with AN	CB	100				2.55		yes		yes							Au		50	Spiro spincoating like Snaith&Grätzel06_2
		6		rinsed with AN	CB	100						yes		yes							Au		50	Spiro spincoating like Snaith&Grätzel06_2
		6		rinsed with AN	CB	100						yes		yes							Au		50	Spiro spincoating like Snaith&Grätzel06_2
0.3	sat. acid in CB	5	RT		CB	100				0.17		0.21	0.1	0.00011 ?	0.06						Au		50	Spiro spincoating after 1 min. of penetrating
										0.17		13	7.6	0.11	64.71									

1 U. Bach, D. Lupo, P. Comte, J. E. Moser, F. Weissörtel, J. Salbeck, H. Spreitzer, M. Grätzel. *Nature*, 1998, **395**, 583-585
2 G. Boschloo, L. Häggman, A. Hagfeldt, *J. Phys. Chem. B*, 2006, **110**, 13144-13150
3 P. Chen, J. H. Yum, F. De Angelis, E. Mosconi, S. Fantacci, S.-J. Moon, R. H. Baker, J. Ko, M. K. Nazeeruddin, M. Grätzel, *Nano Lett.* 2009, **9**, 2487-2492
4 W. M. Campbell, K. W. Jolley, P. Wagner, K. Wagner, P. J. Walsh, K. C. Gordon, L. Schmidt-Mende, M. K. Nazeeruddin, Q. Wang, M. Grätzel, D. L. Officer, *J. Phys. Chem. C.* 2007, **111**, 11760-11762
5 I.-K. Ding, N. Tétreault, J. Brillet, B.E. Hardin, E. H. Smith, S. J. Rosenthal, F. Sauvage, M. Grätzel, M. D. McGehee, *Adv. Funct. Mater.* 2009, **19**, 2431-2436
6 F. Fabregat-Santiago, J. Bisquert, E. Palomares, S. A. Haque, J. R. Durrant, *J. Appl. Phys.* 2006, 100, 034510
7 F. Fabregat-Santiago, J. Bisquert, L. Cevey, P. Chen, M. Wang, S. M. Zakeeruddin, M. Grätzel, *J. Am. Chem. Soc.* 2009, **131**, 558-562
8 D. P. Hagberg, J.-H. Yum, H. J. Lee, F. De Angelis, T. Marinado, M. K. Karlsson, R. Humphry-Baker, L. Sun, A. Hagfeldt, M. Grätzel, M. K. Nazeeruddin, *J. Am. Chem. Soc.* 2008, **130**, 6259-6266
9 T. Horiuchi, H. Miura, S. Uchida, *Chem. Commun.* 2003, 3036-3037
10 W. H. Howie, J. E. Harris, J. R. Jennings, L. M. Peter, *Sol. Energ. Mat. Sol. Cells*, 2007, **91**, 424-426
11 S. Ito, P. Liska, P. Comte, R. Charvet, P. Péchy, U. Bach, L. Schmidt-Mende, S. M. Zakeeruddin, A. Kay, M. Grätzel, M. K. Nazeeruddin, *Chem. Commun.* 2005, 4351-4353
12 S. Ito, S. M. Zakeeruddin, R. Humphry-Baker, P. Liska, R. Charvet, P. Comte, M. K. Nazeeruddin, P. Péchy, M. Takata, H. Miura, S. Uchida, M. Grätzel, *Adv. Mater.* 2006, **18**, 1202-1205
13 K.-J. Jiang, N. Masaki, J.-b. Xia, S. Noda, S. Yanagida, *Chem. Commun.* 2006, 2460-2462
14 K. Kakiuchi, E. Hosono, S. Fujihara, *J. Photochem. Photobiol. A*, 2006, **179**, 81-86
15 C. S. Karthikeyan, K. Peter, H. Wietasch, M. Thelakkat, *Sol. Energ. Mat. Sol. Cells*, 2007, **91**, 432 - 439
16 S. E. Koops, B. C. O'Regan, P. R. F. Barnes, J. R. Durrant, *J. Am. Chem. Soc.* 2009, **131**, 4808-4818
17 J. Krüger, R. Plass, L. Cevey, M. Piccirelli, M. Grätzel, *Appl. Phys. Lett.* 2001, **79**, 2085-2087
18 J. Krüger, R. Plass, M. Grätzel, *Appl. Phys. Lett.* 2002, **81**, 367-369
19 D. Kuang, C. Klein, H. J. Snaith, R. Humphry-Baker, S. M. Zakeeruddin, M. Grätzel, *Inorg. Chim. Acta*, 2008, **361**, 699-706
20 A. B. F. Martinson, J. W. Elam, J. T. Hupp, M. J. Pellin, *Nano Lett.* 2007, **7**, 2183-2187
21 A. J. Mozer, P. Wagner, D. L. Officer, G. G. Wallace, W. M. Campbell, M. Miyashita, K. Sunaharac, S. Mori, *Chem. Commun.* 2008, 4741-4743
22 M. K. Nazeeruddin, T. Bessho, L. Cevey, S. Ito, C. Klein, F. De Angelis, S. Fantacci, P. Comte, P. Liska, H. Imai, M. Grätzel, *J. Photochem. Photobiol.* 2007, **185**, 331-337
23 B. C. O'Regan, J. R. Durrant, P. M. Sommeling, N. J. Bakker, *J. Phys. Chem. C.* 2007, **111**, 14001-14010
24 M. Saito, S. Fujihara, *Energy Environ. Sci.* 2008, **1**, 280-283
25 L. Schmidt-Mende, W. M. Campbell, Q. Wang, K. W. Jolley, D. L. Officer, M. K. Nazeeruddin, M. Grätzel, *ChemPhysChem*, 2005, **6**, 1253-1258
26 L. Schmidt-Mende, S. M. Zakeeruddin, M. Grätzel, *Appl. Phys. Lett.* 2005, **86**, 013504_1-3
27 L. Schmidt-Mende, U. Bach, R. Humphry-Baker, T. Horiuchi, H. Muira, S. Ito, S. Uchida, M. Grätzel, *Adv. Mater.* 2005, **17**, 813-815
28 L. Schmidt-Mende, M. Grätzel, *Thin Solid Films*, 2006, **500**, 296-301
29 H. J. Snaith, L. Schmidt-Mende, M. Grätzel, *Phys. Rev. B*, 2006, **74**, 045306 1-6
30 H. J. Snaith, R. Humphry-Baker, P. Chen, I. Cesar, S. M. Zakeeruddin, M. Grätzel, *Nanotechnology*, 2008, **19**, 424003, 1-12
31 P. Wang, S. M. Zakeeruddin, P. Comte, R. Charvet, R. Humphry-Baker, M. Grätzel, *J. Phys. Chem. B*, 2003, **107**, 14336-14341
32 M. Wang, M. Xu, D. Shi, R. Li, F. Gao, G. Zhang, Z. Yi, R. Humphry-Baker, P. Wang, S. M. Zakeeruddin, M. Grätzel, *Adv. Mater.* 2008, **20**, 4460-4463
33 J.-H. Yum, D. P. Hagberg, S.-J. Moon, M. K. Karlsson, T. Marinado, L. Sun, A. Hagfeldt, M. K. Nazeeruddin, M. Grätzel, *Angew. Chem.* 2009, **121**, 1604-1608

Curriculum Vitae

1978 Birth in Bielawa, Poland
1984 Moving to Langen, Germany
1995 Finished *Mittlere Reife* in the gymnasium Dreieichschule in Langen, Germany
1998 *Abitur* in the technical gymnasium Max-Eyth-Schule in Dreieich-Sprendlingen, Germany
1999 Start of materials science studies at the Technische Universität Darmstadt, Germany
2004 Internship at the Rensselaer Polytechnic Institute (RPI, group of Prof. P. Ajayan) in Troy, NY, USA
2005 Diploma of material science at the TU Darmstadt, Germany

P. L. Marek, *Biomimetic Dye Aggregate Solar Cells*,
Springer Theses, DOI: 10.1007/978-3-319-00636-9,

MIX
Papier aus verantwortungsvollen Quellen
Paper from responsible sources
FSC® C105338
FSC
www.fsc.org

If you have any concerns about our products,
you can contact us on
ProductSafety@springernature.com

In case Publisher is established outside the EU,
the EU authorized representative is:
Springer Nature Customer Service Center GmbH
Europaplatz 3, 69115 Heidelberg, Germany

Printed by Libri Plureos GmbH
in Hamburg, Germany